The History of Stainless Steel

Harold M. Cobb

ASM International®
Materials Park, Ohio 44073-0002
www.asminternational.org

First printing, June 2010

Comments, criticisms, and suggestions are invited, and should be forwarded to ASM International.

Prepared under the direction of the ASM International Technical Book Committee (2009–2010), Michael J. Pfeifer, Chair.

ASM International staff who worked on this project include Scott Henry, Senior Manager, Product Development; Steven R. Lampman, Technical Editor; Ann Britton, Editorial Assistant; Bonnie Sanders, Manager of Production; Madrid Tramble, Senior Production Coordinator; and Patricia Conti, Production Coordinator.

Library of Congress Control Number: 2010921043
ISBN-13: 978-1-61503-011-8 (hard cover)
ISBN-10: 0-61503-011-5 (hard cover)
ISBN-13: 978-1-61503-010-1 (soft cover)
ISBN-10: 0-61503-010-7 (soft cover)
SAN: 204-7586

ASM International®
Materials Park, OH 44073-0002
www.asminternational.org

Printed in the United States of America

The History of Stainless Steel
is dedicated to my dear wife
Joan Inman Cobb

Front Cover

The Chrysler Building, erected in New York City in 1930, was once the tallest building in the world, being almost twice as high as the Washington Monument. It is widely acclaimed as the finest skyscraper, with its art deco style and the ornate tower that is clad with stainless steel.

The Chrysler Building was the first major use of stainless steel in architecture. The Nirosta chromium-nickel alloy had first been introduced in America just three years earlier, and the long-term endurance of the metal in the atmosphere was unknown. The building has become an icon of the stainless steel industry, a symbol of endurance and beauty, and a favorite of architects.

The photograph was taken by Ms. Catherine M. Houska, TMR Stainless, Pittsburgh, Pennsylvania, for the Nickel Development Association, Toronto, Ontario, Canada.

Inside Front Cover

1934 photograph of the Burlington Zephyr at the E.G. Budd Manufacturing Company in Philadelphia, Pennsylvania. Courtesy of the Hagley Museum

Inside Back Cover

List of stainless steels given in Carl Zapffe's 1949 book, *Stainless Steels.*

Back Cover

Top. At a height of 630 feet, the Gateway Arch in St. Louis, Missouri, is the world's tallest monument, which surpassed the 555 foot height of the Washington Monument. With an exterior of stainless steel, the shape of the arch is that of an inverted catenary (or the shape of a chain dangling from two points at the same level). Courtesy of the Jefferson National Expansion Memorial National Park Service, St. Louis, Missouri.

Bottom. The Ford Tudor, one of six Ford Deluxe sedans manufactured by Allegheny Ludlum in 1935 to demonstrate the formability of 18-8 stainless steel and to show its beauty.

Contents

List of Tables and Figures

Tables

Captions for Numbered Figures

Caption for Unnumbered Figure in Chapter 6

Captions for Photographic Insert

Preface

What is stainless steel? The average person has no inkling, but it is all around us, and readers will be surprised to learn some of the stories of this remarkable material that one prominent metallurgist called "the miracle metal."

Every day, most of us use stainless steel tableware and wear a wristwatch with a stainless steel case and band. There are stainless steel racks in refrigerators and ovens, and there are stainless steel toasters, tea kettles, and even kitchen sinks. Cars have stainless steel exhaust systems that last for ten years instead of three when they were made of ordinary steel.

The amazing story is told of Harry Brearley, who rose from poverty, became a self-taught metallurgist, was one of the early discoverers of stainless steel, and received the Bessemer Gold Medal.

In the early days of stainless steel, the metal was often used when the goal was to produce the finest, the most durable, and the most beautiful product that money could buy. The Rolls-Royce Motor Car Company, for example, was one of the first to use stainless steel on an automobile. Their 1929 car displayed the most striking radiator grille imaginable in silvery stainless steel.

In America in 1930, the office building of automaker Walter P. Chrysler opened in New York City. The Chrysler Building was the tallest and most ornate skyscraper in the world. The top 100 feet of the tower was clad in Nirosta stainless steel, making it the most beautiful and most visible building on the New York City skyline.

In 1934, a Philadelphia autobody company tried their hand at building a stainless steel train for the Chicago, Burlington, & Quincy Railroad. It was a streamlined, lightweight, luxurious, silvery train that became the world's fastest. It traveled 3.2 million miles in 25 years and is now on display at the Chicago Museum of Science and Industry.

Eero Saarinen designed the St. Louis Gateway Arch, which was completed in 1965. The 630 foot, stainless-clad arch is the tallest monument. Saarinen wanted the arch to last for a thousand years.

Stainless steel was an expensive material, costing as much as 15 times that of ordinary steel. The story is told of how one young metallurgist in 1970 discovered, in the laboratory, a process that would cut the cost of stainless steel in half and produce better steel. The other part of the story was that it took 12 years to discover how to develop the process for large-scale production.

How it was possible for things like these to happen and the story of how stainless steels were discovered are explained in this first history of stainless steel. Stainless steels have become the third most widely used metals, following aluminum and steel.

Harold M. Cobb
Kennett Square, Pennsylvania
March 2009

Acknowledgments

The author wishes to acknowledge the kind assistance of many individuals and organizations that have been most helpful over a ten-year period in compiling *The History of Stainless Steel.*

Many thanks to Harry W. Weisheit, retired, The Budd Railcar Division, for files of the Railcar Division of the E.G. Budd Manufacturing Co., now of Lansdale, Pennsylvania; R. David Thomas, deceased, former President of Arcos Corp., Philadelphia, Pennsylvania; James D. Redmond, Technical Marketing Resources, Inc., Pittsburgh, Pennsylvania; Ronald Bailey, Plate Division, Allegheny Technologies, Brackenridge, Pennsylvania; Harry E. Lunt, deceased, Burns & Roe, Mendham, New Jersey; and Hubert Langehenke, DIN VDEh, Dusseldorf, Germany.

Many thanks to Alan Harrison, Roger L. Crookes, and David Humphreys, Stainless Steel Advisory Service of the British Stainless Steel Association (BSSA), Sheffield, United Kingdom; William J. Schumacher, A-K Steel Corporation, Middletown, Ohio; Matti Paju, AvestaPolarit, Sweden; Susan Scott, Hotel Savoy, London; David Gymburch, Oneida Ltd., Oneida, New York; The Franklin Institute, Philadelphia, Pennsylvania; The Hagley Museum, Wilmington, Delaware; Valerie Parr, the Kelham Island Industrial Museum, Sheffield, United Kingdom; Louise Fairweather, Outokumpu, Sheffield, United Kingdom; Margaret Lawler, American Society for Testing and Materials International (ASTM), W. Conshohocken, Pennsylvania; and Eleanor Baldwin, ASM International, Materials Park, Ohio. My sincere appreciation also to Catherine M. Houska, TMR Stainless, Pittsburgh, Pennsylvania; Gary E. Coates, Nickel Institute, Toronto, Canada; Evelyn D. Roberts, Pittsfield, New Hampshire; and Kathleen

Moenster, Librarian, Jefferson National Expansion Park, St. Louis, Missouri.

Special thanks to Sonia S. Ralston, Kennett Square, Pennsylvania. The author is indebted to John P. Moran, retired, G.O. Carlson Co., Burlingame, California; Brian McCarthy, President of the Flying Yankee Restoration, Lincoln, New Hampshire; Karl G. Reed, retired, Aviation Division, E.G. Budd Manufacturing Company, Kennett Square, Pennsylvania; and Richard Blanchard of Kennett Square, Pennsylvania. The author thanks Steve Lampman of the ASM International staff for his guidance and for shepherding the work through to publication. And last but not least, many thanks to my wife, Joan I. Cobb, for proofreading the manuscript and for her many suggestions.

Recognition is given to Outokumpu (the successor company to British Steel Stainless), Sheffield, United Kingdom, for granting permission to reprint portions of *Harry Brearley—Stainless Pioneer*.

Harold M. Cobb
Kennett Square, Pennsylvania
October 2009

Credits

The author gratefully acknowledges the following persons and organizations that have given permission to use illustrations and other materials in *The History of Stainless Steel.* Acknowledgments and permissions for figures are cited in the captions.

- American Iron and Steel Institute
- American Society for Testing Materials, 1924
- Brian McCarthy, President, Flying Yankee Restoration Group, Inc.
- Catherine M. Houska, TMR Stainless
- Craig Clauser, Craig Clauser Engineering Consulting Incorporated
- D. Gymburch
- Elwood Haynes Museum, Kokomo, Indiana
- Franklin Institute Museum, Philadelphia, Pennsylvania
- Hagley Museum and Library, Wilmington, Delaware
- J&L Specialty Steel
- Jefferson Expansion Memorial National Park
- Kearns Communications Group
- Louise Fairweather, Outokumpu-Sheffield (successor company to British Steel Stainless), in Chapter 5 for use of excerpts from *Harry Brearley—Stainless Pioneer*, published by British Steel Stainless, 1988
- Nickel Development Institute
- Pittsburgh Civic Arena
- Princeton Architectural Press
- Sheffield Industrial Museums Trust
- Zapffe family

About the Author

Harold M. Cobb graduated from Yale University in 1942, receiving a B.E. degree in metallurgical engineering. He has had a broad background in the stainless steel industry, where he was involved in the development of new stainless steel products, including watch screws, hollow stainless steel aircraft propeller blades, roll-formed compressor blades and vanes for jet engines, boron carbide stainless steel for moderating nuclear reactors, and sinter-bonded porous stainless steel fiber-metal products.

Cobb's industrial experience included positions at the Edward G. Budd Manufacturing Co., Westinghouse Aviation Gas Turbine Division, United Nuclear Corp., and as chief metallurgist at Clevite Aeroproducts and Pratt & Whitney.

He was chairman of the Philadelphia and Connecticut sections of the American Institute of Mining, Metallurgical and Petroleum Engineers (AIME). He holds a patent on a manufacturing process for nuclear fuel elements.

After 22 years in the metals industry, Cobb became a manager at the American Society for Testing and Materials (ASTM) in Philadelphia, working with many of the metals technical committees, including Committee A-10 on Stainless Steel. He was one of the principal promoters and developers of the Unified Numbering System (UNS) for metals, which was organized jointly by the Society of Automotive Engineers (SAE) and ASTM in 1970. For many years, Cobb developed and served as the number assigner for the miscellaneous steels series of UNS numbers, the K series. He has been the principal editorial consultant for the last four editions of *Metals and Alloys in the Unified Numbering System (UNS)*.

Cobb served as Secretary of the U.S. Secretariat for the International Standards Committee ISO/TC17/SC12 on Carbon Steel Sheet and Strip for 15 years. He has edited 22 books on steel, including works on carbon, alloy and coated steel sheet and strip, tool steels, stainless steel specifications, and a *Pocketbook of Standard Wrought Steels*. In 1999, he became editor of the *Stainless Steels Products Manual*, one of the 16 steel products manuals that the American Iron and Steel Institute (AISI) initiated in the 1950s. In 2008, Cobb edited and substantially revised his second edition of *Stainless Steels*, now published by the Association for Iron and Steel Technology.

He has written the articles "Development of the Unified Numbering System for Metals," "The Naming and Numbering of Stainless Steels," and "The 75th Anniversary of the Burlington Zephyr Stainless Steel Train." Cobb is a member of ASTM Committee A-1 on Steel, Stainless Steel, and Related Alloys and is a Life Member of ASM International.

The History of Stainless Steel
Harold M. Cobb

CHAPTER 1

Introduction

"Starting from rust, men have produced something which looks like platinum and resists chemical attack like gold, and yet a square inch can support a quarter of a million pounds . . . this is the crowning achievement of metallurgy."
Stainless Steel—The Miracle Metal
Carl Zapffe, metallurgical consultant, 1960

AS LATE AS the year 1910, the following statement appeared in the British journal *The Corrosion and Preservation of Iron and Steel* (Cushman and Gardner): "The tendency to rust is a characteristic inherent in the element known as iron, and will, in all probability, never be totally overcome."

As the first hundred years of what may well be called "The Stainless Steel Age" draws to a close, it seems an appropriate time to tell the story of the remarkable discoveries of stainless steels and their myriad applications.

There were quite a number of nonferrous metals that were available for use under certain corrosive conditions. They included nickel, nickel silver (the nickel-copper-tin alloy that contained no silver but resembled it somewhat), copper, brass, and bronze. There were also two new alloys, aluminum and Monel, a nickel-copper alloy discovered in 1905 that was used for roofing the new Pennsylvania Railroad Station in New York City.

The nonferrous metals served well, but they were more expensive than steel and not as strong. As a result, there was extensive use of ordinary steels with coatings to resist corrosion, coatings that often consisted of nonferrous metals such as zinc and tin, which could be applied by dipping the steel into the molten metals or by electroplating the steel with copper, brass, tin, nickel, chromium, or zinc. Painting, of course, was frequently used. Sinks were mostly made of heavy, porcelain-enameled cast iron. The life of the coated steels depended on the thickness of the applied coating, the quality of the work, and the environment in which the coated material was used. Unless the coated steel products were used indoors, the coating eventually broke down over a period of years, usually by cracking or pitting, which exposed bare steel.

In the Middle Ages, alchemists tried to turn lead into gold, but no thought had ever been given to the possibility of turning iron into a rustproof noble metal, nor were any rewards offered for such an accomplishment. Iron rusts, and that is the nature of it.

In the 19th century, at least 25 scientists, working in England, France, Germany, and the United States, were conducting experiments and writing papers about alloys of iron and varying amounts of chromium, nickel, and carbon. It was generally noticed that alloys with more chromium were somewhat more resistant to corrosion in many environments than carbon steel, but not a single person had experimented with an alloy having chromium and carbon contents similar to the alloys that would become known as stainless steels.

However, Robert A. Hadfield had created samples with up to 9.18% chromium. Because he had tested his samples in sulfuric acid, they all dissolved. He concluded that chromium decreases the corrosion resistance of steel, and he discontinued his experiments with chromium. Chromium-containing alloys that we now call stainless steel are not resistant to sulfuric acid.

Seven men, unknown to each other and living in four different countries, inadvertently discovered alloys that we now call stainless steel in the period from approximately 1905 to 1912. The discoverers thought that alloys with greater corrosion resistance must have some useful applications, but it was only a guess. There were also quite a few difficulties: The alloys were quite expensive, as compared to common steel, and quite a lot of the chromium was lost in the slag during melting. In fact, practically every step of the manufacturing process was different from ordinary steelmaking and usually more difficult. One book that was written some years after the discoveries advised that "melting should be performed with care" and that "forging was difficult," cautioning that "the billets are best cooled on the floor out of draughts." The scale needed to be removed by grinding, because pickling did not work with these alloys. However, grinding had to be performed carefully to avoid the formation of hot spots that would lead to cracking. Machining was particularly difficult.

Despite the problems mentioned, there were some notable successes in finding buyers for the alloys. After several tries, Harry Brearley eventually found a cutler in Sheffield who succeeded in making some fine knife blades out of Firth's iron-chromium alloy. There was a brief demand for the alloy in the cutlery industry, and a short announcement of the development appeared in the *New York Times*. Use of the alloy for knifemaking came to a halt with the beginning of World War I, but the Royal Air Force had found that the cutlery steel was just the ticket for making "aeroplane" exhaust valves and immediately ordered that every pound of the metal Firth could make should be shipped to the aircraft engine factories. Firth soon began selling the metal under the name Firth's Aeroplane Steel, or FAS, and the stainless steel business was indeed off to a flying start.

The Krupp Works in Germany had developed an iron-chromium-nickel alloy that was highly resistant to acids. They soon made large sales of the metal to a chemical company for building nitric acid storage tanks.

A great new industry had been launched that, within 20 years, would produce 60,000 tons of the alloys a year in America alone.

What to call the alloys was something of a problem, because each of the three original classes needed to be named. The alloy for cutlery was called, appropriately enough, cutlery steel or rustless steel. Legend has it, though, that the cutler who made the first successful knife blades of the alloy was heard to say, "It stains less." Firth then started calling it stainless steel, and the cutlers began proudly stamping their knife blades with those words.

The relatively soft iron-chromium alloy that, unlike the cutlery steel, could not be hardened by heat treatment became known as rustless iron or stainless iron.

In Germany, the Krupp Steel Works called their alloy that was remarkably resistant to most acids iron-chromium-nickel corrosion-resisting alloy or, probably more often, Krupp V2A alloy. It was the custom of each producer to give every different alloy a name, especially one that they thought customers might easily remember. Names cropped up such as Anka, Staybrite, and Vesuvius in England and Enduro S, No-Kor-O-H, and Circle L No. 19 in America.

In the 1920s, metallurgists in this business decided that the naming of these alloy classes should be on a more scientific basis, which was according to their crystal structure. When a sample of a metal is polished and etched, the crystal structure can be determined when the sample is examined with the aid of a high-powered microscope.

The crystals of the cutlery steels are recognized by their shape as what metallurgists call martensite, and the alloys in this class can all be called martensitic alloys. The crystals of the class described earlier as stainless irons are known as ferrite, and the alloys are called ferritic alloys. For the third class, which is the iron-chromium-nickel corrosion-resisting alloy, the crystals are austenite, and the alloys are called austenitic alloys. These three names—martensitic, ferritic, and austenitic—are used frequently throughout this book to identify the alloy class under discussion.

For quite a few years, only the cutlery alloys were called stainless steel, the name originally adopted by Firth. This made sense, because these were the only alloys that had similar characteristics to low-alloy steel, with the exception that they did not rust. The popular names for the other two classes became stainless iron and nonrusting chromium-nickel alloy. The committee established by the American Society for Testing Materials (ASTM) in 1929 was named Committee A-10 on Corrosion-Resistant and Heat-Resistant Alloys.

In 1933, when the first major publication in America was written about these alloys, Ernest Thum called his book *Stainless Steels* and gave his rationale for this title in the introduction:

> "Probably most attentive readers of these first few pages have sensed the difficulties which arise with respect to the classification and nomenclature of high chromium alloys. Especially is the latter badly cluttered up with trade names and loose terminology. Even though laying himself open to the charge of taking the line of least resistance, the Editor has grouped all the high chromium heat and corrosion resisting steels under the term 'stainless steel.' Few of them are really stainless, many of them are not steels in the sense that they do not harden very much during quenching. Likewise, an attempt has been made by commercial organizations in England to restrict 'stainless steels' to the cutlery types, and in America to high chromium-iron alloys. Nevertheless, common usage, which grows with a fine disregard of industrialists, scientists and even grammarians, has broadened the term to include all steely colored metals that do not tarnish readily. We might as well bow to usage!"

Thum's classic work eventually had almost everyone in the industry calling all of these high-chromium alloys stainless steels. The usage caught on worldwide, being changed only slightly in the translation. In France, for example, it is *acier inoxydable* (nonoxidizing steel), and in Germany it is *nichtrostende Stahl* (nonrusting steel).

How is stainless steel defined? Until recently, dictionaries that included a definition for stainless steel defined the term approximately as follows: "An alloy of iron containing at least four percent of chromium and having good corrosion resistance." This definition is unfortunate because it should state "at least 10.5 percent of chromium," which is the percentage at which there is a dramatic increase in the corrosion resistance. With only 4% chromium, the alloy has enough corrosion resistance for only some mildly corrosive applications in oil refining. How the dictionaries got started with the 4% chromium definition is a fairly long story that is explained in Chapter 16, "The Naming and Numbering of Stainless Steels."

This book covers a broad spectrum of historical events, many of which have not been touched upon in other works on stainless steel. It includes the discoveries of the various metallic elements that are used in the various alloys of stainless steel and discusses numerous experiments conducted during the 19th century with iron-base alloys containing chromium and carbon. The actual discoveries of stainless steel during the period from 1905 to 1912 are discussed, and the very important beginnings of commercial production in approximately 1913 are covered in greater detail.

In addition to the discovery of the alloys, the book recounts events important in the overall history of stainless steel, including the first great meeting in 1924 that was attended by virtually all of America's stainless steel producers as well as Dr. Benno Strauss, the German inventor of stainless steel. Also discussed are the important books and journals, the steelmaking processes, why stainless steel is stainless, the numbering and naming systems, the greatest promoters of stainless steel, the committees and associations, the standards and specifications, the discovery of the precipitation-hardening grades and the duplex alloys, the invention in 1970 of a melting and refining system that improved the quality and slashed the cost of stainless steel by half, the dates when many stainless steel products entered the marketplace, and some interesting stories surrounding the use of stainless steel in architecture, trains, automobiles, the aerospace industry, cookware, and tableware.

A final section, Appendix 2, "A Stainless Steel Timeline," lists 460 events of interest concerning stainless steel.

CHAPTER 2

The Early Discoveries

STAINLESS STEEL is a remarkable achievement of modern metallurgy. It has been described as "the miracle metal" and "the crowning achievement of metallurgy" by the prominent 20th century metallurgist Carl Zapffe. Like many technological innovations, the history of stainless steel is built on the shoulders of many, with breakthroughs from individuals with vision, persistence, and luck. Stainless steel is not only a shining achievement of modern metallurgy but also an enduring legacy of human industry and progress.

The early discoveries occurred in the 18th and 19th centuries, beginning with the identification of chromium as an element. Iron-chromium alloys were also experimentally investigated during the 19th century. The significant surge of advancement happened in the early part of the 20th century with a series of discoveries made within the context of the commercial use of iron-chromium and iron-chromium-nickel alloys and the emergence of metallurgy as a modern engineering discipline.

The early discoverers and pioneers of stainless steel were men of both science and industry, and some brief historical descriptions of stainless steel occurred in the decades after commercial development. One early account was a brief one-page history in the trade booklet *Firth-Sterling 'S-Less' Stainless Steel* (Firth-Sterling Steel Co., McKeesport, Pennsylvania, 1923), with technical information on stainless steels (including terms of warranty to replace "any steel which proves defective when properly treated and used for the purposes specified in the order . . ."). Another book, *Stainless Iron and Steel* (Chapman & Hall, Ltd., London, 1926) by J.H.G. Monypenny, contained a chapter on the early history of stainless steel.

Other historical reviews on the development of stainless steels are contained in *The Book of Stainless Steels,* edited by Ernest Thum (The American Society for Steel Treating, 1933), and a chapter by Carl Zapffe in his book *Stainless Steels* (The American Society for Metals, 1949). Zapffe describes six key discoverers who studied the metallurgy of various iron-chromium alloys. Five of the six early discoverers of stainless steel are pictured in Fig. 1. The other discoverer noted by Zapffe is W. Giesen, who in 1909 published a lengthy account in England on austenitic nickel-chromium stainless steel and steels containing 8 to 18% chromium with 0.3% carbon.

Léon Guillet (1873–1946) (Fig. 1) was the first to metallurgically and mechanically investigate chromium and chromium-nickel steels with martensitic, ferritic, and austenitic crystalline structures. However, he did not discover the corrosion resistance of these steels until later, and he is not known to have pursued work on the corrosion resistance of chromium steels. The discovery of what was subsequently known as stainless steel is usually credited to Harry Brearley (1871–1940) (Fig. 1), who produced the first commercial cast of a martensitic stainless steel, which at first was used for cutlery.

The Discovery of Chromium (1797)

In 1797, the French chemist Louis Nicholas Vauquelin (1753–1829) discovered chromium oxide in an ore of "red lead" from Siberia. The following year, he isolated the new metal, chromium, by the heating of chromium oxide and charcoal. He called the metal chrome for the Greek word *chromos*, meaning color, after observing that most of the compounds of chromium are a variety of colors.

The world's first piece of chromium metal was presented to the French Academy by Vauquelin in 1798. Chromium would become the key ingredient of stainless steel.

Michael Faraday Pioneers the Alloying of Steel (1820)

Michael Faraday (1791–1867), who is best known for his pioneering work in the fields of electricity and magnetism, also appears to have been one of the first men to work on the development of alloy steels.

Fig. 1 Five discoverers. Source: Zapffe, 1949

In the paper "Experiments on the Alloys of Steel, Made with a View to Its Improvements," Stodart and Faraday stated that the aim of their work was twofold:

- To ascertain whether better cutting instruments may be produced
- To ascertain whether any such alloys would, under similar circumstance, prove less susceptible to oxidation

A collateral object of research was to discover new metallic materials for reflecting mirrors.

Their research covered the addition to steel of noble metals, including silver, platinum, and rhodium, as well as nickel and chromium. While the alloy containing 10% nickel was found to be noticeably more resistant to tarnishing when exposed to air, the best corrosion resistance was found in alloys with platinum and rhodium containing 50% or more of the alloying elements—clearly not practical possibilities. The highest chromium content they tried was only 3%. This conferred excellent cutting properties but no improvement in corrosion resistance.

Iron-Chromium Alloys and the Production of Ferrochromium (1821)

In 1821, which was already a quarter-century after Vauquelin's discovery of chromium metal, the Frenchman Pierre Berthier (1782–1881) found that iron highly alloyed with chromium was more resistant to acids than chromium-free iron and that the resistance increased with increasing chromium content. Berthier's attention had been drawn to a publication in England by Stodart and Faraday the year before, which described one of the first attempts to add chromium to steel. Berthier proceeded to produce iron-chromium alloys directly by reduction of the combined oxides, thus obtaining what is now called ferrochromium.

Producing ferrochromium was a historic event, because it can be added to molten iron to produce stainless steel (containing 10 to 30% chromium). Berthier's ferrochromium contained 17 to 60% chromium and had a very high carbon content. Although not realized at the time, this high carbon content would preclude the possibility of producing stainless steel.

Although Berthier did produce a steel containing chromium, only a low amount of that element (1.0 to 1.5%) was present. He made a knife and a razor blade of fine quality from it and observed the greater hardness of the new steel. Even more remarkable, in view of later events, he commented on the improved corrosion resistance of the steel and recommended chromium steel for cutlery. However, Berthier worked only with steels of high carbon content and with chromium contents either too high or too low to fall within the range later discovered to have the unique qualities of stainless steel. Another 90 years elapsed before stainless steel was truly discovered.

Woods and Clark Describe an Acid- and Weather-Resistant Alloy (1872)

Two Englishmen, J.E.T. Woods and J. Clark, discovered an acid- and weather-resistant alloy that they described as follows:

> "The alloy we prefer to use for anti-acid metal consists of 5% tungsten and 95% chromium combined in the proportion of 33% of alloy to 67% of steel. This metal also is very hard, is of a silvery colour, takes a high polish which it retains in a damp or oxidising atmosphere and is accordingly extremely useful for various purposes where high reflective qualities are required, various parts of instruments where German Silver is now used, for coinage metal and for cutlery which has to be used in contact with acids."

In June 1872, they obtained Provisional British Patent 1932, but the final specification was never filed, and there seems to be no record of further work undertaken on this or similar alloys.

Discoveries in the 1890s

High-Chromium Steels Investigated (1892). In England, Sir Robert A. Hadfield (1858–1940), the inventor of Hadfield's manganese steel, investigated the corrosion resistance of steels containing up to 16.74% chromium and high carbon contents in the range of 1 to 2%. He tested his samples with a solution of 50% sulfuric acid under the erroneous assumption, common at the time, that corrosion testing in sulfuric acid was representative of corrosion in general and would serve as an accelerated test.

As a result, Hadfield reached the conclusion that chromium actually impairs the corrosion resistance of steel. His unfavorable and unfortunate conclusions threatened to curtail further interest in high-chromium steels.

It is interesting to note that Hadfield's invention of Hadfield's manganese steel in 1882 was the first very high-alloy steel of commercial importance. His alloy contained 12% manganese and was austenitic. It could not be hardened by heat treatment, but it rusted similar to ordinary steel. The alloy was highly abrasion resistant and was used in mining and railroad equipment.

The Discovery of Low-Carbon Ferrochromium (1895). A German chemist, Hans Goldschmidt (1861–1923), discovered a method for producing low-carbon ferrochromium and chromium metal by aluminothermic reduction. The availability of low-carbon ferrochromium, although quite expensive, made it possible to produce chromium steels with reduced carbon contents for the first time. This important event soon led to the development of the first stainless steels by metallurgists in England, France, Germany, and the United States.

The Discovery of the Effect of Carbon on the Corrosion Resistance of Iron-Chromium Alloys (1898). In France, A. Carnot and E. Goutal reported their discovery that high carbon contents have a damaging effect on the corrosion resistance of iron-chromium alloys.

The Discovery of Martensitic and Ferritic Chromium Stainless Steels (1904)

Beginning in 1904, Léon Guillet (1873–1946), professor of metallurgy and metal processing at the Conservatoire des Arts et Métiers, published a series of research papers on iron-chromium alloys having carbon contents acceptably low for modern stainless steel analyses. He made his steels with carbon-free Goldschmidt chromium (oxide ore reduced to chromium with powdered aluminum) and controlled the carbon contents so that they ranged from 0.043% in low-chromium alloys and from 0.142% in alloys of stainless composition up to 1.0%. Among these alloys were steels falling within the range of modern analyses for martensitic and ferritic stainless steels. The martensitic, hardenable steels would later be described as AISI types 410, 420, and 440C, and the ferritic nonhardenable steels would later be described as AISI types 442 and 446.

Guillet studied these steels with great thoroughness for the tests available at that time, describing their fundamental metallographic features, heat treatment, and mechanical properties. He also identified their structures as martensite or ferrite. However, he did not discover until later the phenomenon of passivity in these steels, which sets them apart from the standpoint of corrosion resistance. Guillet is not known to have pursued work on stainless corrosion resistance, nor did he obtain patents.

The Discovery of the Chromium-Nickel Austenitic Stainless Steels (1906)

In addition to his study of martensitic and ferritic stainless steels, Léon Guillet published a detailed study of iron-chromium-nickel alloys that fall into a third important group of stainless steels that are austenitic in structure. His chemical compositions were not strictly those defined today as preferred grades, but they were of such similar constitution that the only difference lies in refinements adopted many years later. Guillet therefore seems to be the first to metallurgically and mechanically explore the stainless steels, and he did this for the three important subdivisions of these alloys—martensitic, ferritic, and austenitic. Several years after the first austenitic stainless steel was made by Guillet and Geisen, commercial austenitic stainless steel was developed by the joint research of Eduard Maurer and Benno Strauss at the Krupp Research laboratories from 1909 to 1912. In 1908, prior to his research with Strauss at Krupp, Maurer worked on his doctorate under Osmond on the heat treatment of high-chromium steels. Maurer discovered proper heat treatments.

The Discovery of Corrosion Resistance (1908)

In 1908 in Germany, Philip Monnartz (Fig. 1) studied the effect of carbon content on the corrosion resistance of high-chromium steels. His research showed the stainlessness of stainless steel as a function of the passivity phenomenon, and in 1911 he published a brilliant piece of work on this subject in *Metallurgie*, entitled "The Study of Iron-Chromium Alloys with Special Consideration of Their Resistance to Acids." The paper presented the following conclusions:

- "The stainlessness of chromium steel is disclosed by a precipitous drop in the corrosion rate when the alloy content nears 12 percent chromium, particularly when tested in nitric acid, but also in other acids, waters and the atmosphere."

- "Passivity is the phenomenon responsible for this great increase in corrosion resistance."
- "Passivation is dependent upon oxidizing conditions, as opposed to reducing conditions."
- "A pre-passivation treatment consists of immersing the steel in nitric acid, or by letting it stand for a period in air, before exposing it, for example, to sulfuric acid in which stainless steels serve undependably."
- "Passivation is best maintained in cold sulfuric acid, least in hot, and when once activated, the steel dissolves rapidly."
- "Low carbon content is important since excess carbon forming free carbides in the alloy acts to prevent passivation."
- "The carbon can be stabilized so that it does not affect the corrosion resistance by adding quantities of other elements such as titanium, vanadium, molybdenum, or tungsten."
- "Molybdenum has an especially favorable effect in enhancing corrosion resistance."

These were remarkable discoveries. Monnartz appears to have been the first to discover the reasons for what he called the "stainlessness." Monnartz is the first to have pointed out the significance of at least 12% chromium in an alloy. He is apparently the first to have used the term *passivity*, which is described as being dependent on having oxidizing conditions. He also says that nitric acid can be used to induce this condition of passivity.

Monnartz also appears to be the first to discover that free carbides, or what we call carbide precipitation, are detrimental. And finally, Monnartz explains how an alloy can be stabilized by adding certain elements such as titanium. He also stresses the value of molybdenum.

Monnartz has shown that he has discovered stainless steel and why it is stainless. His work complements that of Guillet.

Another Important Ferritic Chromium Stainless Steel Is Discovered (1911)

In France, Albert M. Portevin (1880–1962) (Fig.1) studied both martensitic and ferritic alloys and published his observations at length in England in 1909 and in France and America in 1911 and 1912. Portevin used many of Guillet's original alloys for his studies, but he also listed additional alloys, the most important of which appears to be an alloy consisting of 17.38% chromium and 0.12% carbon, which is exactly the prominent type 430 stainless steel that has been used for many years. Portevin's contribution more or less completed the broad discoveries of the principal alloys of the martensitic, ferritic, and austenitic families of stainless steels.

CHAPTER 3

Discoveries of the Commercial Usefulness of Stainless Steel

CHROMIUM STEELS had been used since 1869 for their intense hardness, but Harry Brearley melted the first commercial chromium steel as a stainless steel for cutlery blades in 1913. Contemporaneous developments were also occurring elsewhere in Germany and America. At Krupp laboratories from 1909 to 1912, high-chromium nickel steels were developed by Eduard Maurer (Fig. 1) and Benno Strauss (Fig. 2). During this time, several other pioneers in America were also experimenting with high-chromium steels for potential commercial application. In 1911, Elwood Haynes (Fig. 2) experimented with high-chromium steels to determine the effect of chromium on corrosion resistance, hardness, elasticity, and cutting qualities. Christian Dantsizen (Fig. 2) of The General Electric Research Laboratory also began experimenting in 1911 with high-chromium steels for possible commercial development.

Usefulness of a Martensitic Chromium Stainless Steel Discovered in England and America (1911–1912)

Harry Brearley (Fig. 3) is generally credited above all others as the initiator of the industrial era of stainless steel, and rightly so, at least as far as the hardenable chromium stainless steels are concerned. In 1912, while investigating the development of new alloys for rifle bar-

Fig. 2 Six pioneers. Source: Zapffe, 1949

Fig. 3 Harry Brearley. Source: Copyright. Sheffield Industrial Museums Trust. Reprinted with permission

rels, he noted that the customary etching reagents for steel did not etch steels having high chromium contents, and he concluded that a steel with upward of 10% chromium may be of advantage in this particular application. He then produced an ingot containing approximately 13% chromium and 0.24% carbon. It seems that the alloy did not show the desired improvement with regard to the gun barrels; however, material from the same cast was eventually made into cutlery blades. This, then, was the first commercial cast of what was subsequently known as stainless steel.

Brearley's actual account of the story was as follows:

> "When microscopic observation of these steels was being made, one of the first things noticeable was that the usual reagents would not etch, or etched very slowly, steel containing low carbon and high chromium. I was satisfied that merely to specify the composition of the material was not sufficient, as in different conditions of heat treatment some would etch and some would not.
>
> "The use of the material for ordnance purposes, as originally intended, appeared to excite no interest. But I reported later on the unusual non-corrosive properties to each of the firms with which I was associated, suggesting purposes to which this material might be advantageously used, including cutlery. Nobody was impressed; perhaps the idea of producing on a commercial scale a steel which would not corrode sounded ridiculous."

However, Brearley was not deterred, and finally, in 1914, he met Ernest Stuart, cutlery manager of Messrs. R.F. Mosely's, who believed that a rustless steel could not be made but agreed to make some cheese knives. He first worked on some small samples with varying success, but then, with Brearley's help, forged and heat treated approximately 125 pounds of Firth's Aeroplane Steel, as it was called then, into knives that were found to be just fine. The steel that had been declared dead and almost useless became an absorbing topic of conversation among cutlers and steelmakers and a subject for newspaper articles. A new industry had been born in Sheffield!

Brearley went on to obtain U.S., French and Canadian patents. Because of his diligence, a stainless steel cutlery industry was started in England. He organized the Brearley-Firth Stainless Steel Syndicate and the American Stainless Steel Company (see Chapter 5, "The Life of Harry Brearley (1871–1948)").

Elwood Haynes (1857–1925) (Fig. 2) of Kokomo, Indiana, also deserves mention with respect to the development of martensitic stainless steels. Although better known today for his invention of Stellite alloys, Haynes began making experiments in 1911 to determine the effect of chromium on the resistance of iron and steel to chemicals and the atmosphere. He also investigated the effect of chromium on the hardness, elasticity, and cutting qualities of steel.

Haynes actually filed a patent application a little earlier than Brearley, but it was at first denied because there were already patents for chromium steels. On April 1, 1919, Haynes was finally granted U.S. Patent 1,299,404 for "Improvements in Wrought Metal Articles," which covered steels with 5 to 60% chromium and 0.1 to 1.0% carbon. Because Brearley's U.S. patent granted in 1916 covered similar claims, this resulted in a dispute that was later resolved to the satisfaction of both Haynes and Brearley when they assigned their patents to and became major shareholders of the American Stainless Steel Company, a patent-holding company formed in 1918.

Usefulness of Ferritic Chromium Stainless Steels Discovered in America (1911–1914)

Credit for the development of the nonhardenable, ferritic stainless steels has generally been given to the two Americans, Frederick M. Becket (1875–1942) and Christian Dantsizen (Fig. 2), although these were essentially alloys made earlier by Guillet, Portevin, and Monnartz.

Beginning in 1911 with experiments on lead-in wire for electric bulbs, Dantsizen of The General Electric Research Laboratory developed alloys containing 14 to 16% chromium and 0.07 to 0.15% carbon. Unlike Brearley's chromium alloys, these were not hardenable by heat treatment because of their very low carbon content or their high chromium content. In 1914, the use of Dantsizen's stainless steel was extended to steam turbine blades, for which essentially the same alloy is used today.

The development of the higher-chromium ferritic alloys was due to the work of Frederick Becket of the Electro Metallurgical Company at Niagara Falls, New York. While investigating the effect of chromium on oxidation resistance at 2000 °F (1095 °C), Becket observed a marked increase in resistance as the chromium content was raised above 20%. He produced an oxidation-resistant "chrome-iron" containing 25 to 27% chromium that, in spite of a high carbon content, was not hardenable because the chromium content was too high to permit significant hardening by heat treatment.

Becket is also noted for his 1908 invention of low-carbon ferrochromium produced by reducing the chromium oxide with silicon instead of aluminum, as used in Goldschmidt's process.

Usefulness of Chromium-Nickel Stainless Steels Discovered in Germany (1912)

In 1909, Eduard Maurer (1886–1969) joined the research laboratory at the Friedrich A. Krupp works at Essen as their first metallurgist. In 1912, while searching for materials suitable for use in pyrometer tubes, Maurer discovered that some iron-chromium alloys with approximately 8% nickel, which Benno Strauss had made, were impervious to attack after months of exposure to acid fumes in his laboratory.

Dr. Strauss told the story of his discovery in a paper presented at a symposium held in 1924 in Atlantic City, New Jersey, by the American Society for Testing Materials (ASTM). The following excerpt is from that paper:

> "Metallurgists for a long time have been accustomed to alloy iron with nickel with a view to the improvement of the iron, whereas the discovery of the valuable properties of chromium is of comparatively recent date. In seeking for a suitable material from which to make protective tubes for thermo-electric couples, I started on the supposition that it ought to be possible to make use of the valuable properties of chromium, that is, its resistance to the action of oxygen, for the purpose of increasing the resistance of iron against hot gases. In 1910, five experimental alloys high in chromium were prepared for this purpose. Polished specimens taken from them, after having been exposed for some time to the atmosphere, revealed the curious fact that particularly those steels with 20% chromium had remained perfectly bright, while other bars, even those containing up to 25% nickel, had become rusty.
>
> "As investigations were being carried out at the same time with regard to the corrosion of metals in fresh water and sea water, it was deemed expedient to extend this investigation to both chro-

mium and chromium-nickel steels. As, however, only such alloys as could be put to practical uses were to be selected for the purpose, it became necessary to investigate the tensile strength of these alloys considered as a function of the heat treatment undergone by them. In connection with the study of the mechanical strength of the chromium-nickel steels, their structure and other physical properties were also looked into. This work was carried out in collaboration with Mr. Maurer during 1909 to 1912."

Patents in 1912, and a paper presented by Strauss in 1914 before a meeting of chemists in Bonn, defined these alloys as comprising the following two types:

- 7.0 to 25% chromium, 1.0% maximum carbon, and 0.5 to 20% nickel (German Patent 304,126)
- 15 to 40% chromium, 1.0% maximum carbon, and 4 to 20% nickel (German Patent 304,159)

The first type had to do particularly with a martensitic grade, V1M, containing 14% chromium and 2% nickel, similar to England's Two Score alloy (20% chromium, 2% nickel) and our modern types 414 (12% chromium, 2% nickel) and 431 (16% chromium, 2% nickel), and was recommended for general corrosion resistance.

The second type was an austenitic grade, V2A, which was 20% chromium, 7% nickel, or quite close to the 18-8 of today, and was recommended for its exceptional corrosion resistance, particularly in nitric acid. Articles made from both types of steel were displayed at the Mälmo Exhibition in Sweden in 1914.

Maurer and Strauss are recognized for obtaining patents and establishing the commercial usefulness of the chromium-nickel stainless steels that have now become the most widely used of all the classes of stainless steel.

Usefulness of Chromium-Silicon Steels

As a Ludlum Steel Co. metallurgist, P.A.E. Armstrong (Fig. 2) discovered chromium-silicon steels in 1914, when a small electric furnace was accidentally contaminated due to some silicon reduced from the asbestos cover on the electrode. World War I interrupted his efforts, but he resumed his research after the Armistice. He made no less than 2500 melts to test the properties of his chromium-silicon steel. Armstrong patented chromium-silicon steels in 1919, and they were marketed under the trade names Neva-Stain and Silchrome. The major application was gas engine exhaust valves. Chromium-silicon steels continue to be principally used for exhaust valves.

In 1917, Charles Morris Johnson (Fig. 2) of the Crucible Steel Company began investigating chromium-nickel-silicon steels, which were later patented under the name Rezistal. U.S. Patents 1,420,707 and 1,420,708 were issued in June 1922. Johnson first described these alloys in a paper at the 1920 Philadelphia convention of the American Society for Steel Treating. The alloys were austenitic due to the nickel content, and silicon imparted improved resistance to scaling. With a high amount of chromium, nickel, and silicon content, some alloys had less than 50% iron and thus crossed the boundary of being steel.

CHAPTER 4

The Great Stainless Steel Symposium (1924)

IN 1924, the American Society for Testing Materials (ASTM) organized the symposium "Corrosion and Heat Resisting Alloys, and Electrical Resistance Alloys," which was held June 24 to 27 in Atlantic City, New Jersey. The event was attended by those in the stainless steel business and by representatives of companies that were interested in getting into that business. There were also representatives from the nonferrous metals industry and those companies producing electrical resistance alloys. It was the beginning of a major role that ASTM (later renamed the American Society for Testing and Materials) subsequently played in the history of stainless steel.

ASTM began at the turn of the 20th century, when steam engines and railroads were widely used for moving people and freight. However, there were no standards for iron and steel products, nor were there standard methods of testing those materials. As a result, locomotive and other boiler explosions were common, as were failures of steel track, railroad trestles, and locomotive wheels and axles, usually involving a considerable loss of life and property.

Mansfield Merriman, a graduate of the Sheffield Scientific School at Yale and a professor of civil engineering at Lehigh College, and Edgar Marburg, a professor of civil engineering at the University of Pennsylvania, were among those who held meetings in Philadelphia in 1898 to address these problems. The meetings led to the organization of ASTM and the establishment of a Committee on Steel that began with 60 members. In two years, the committee developed a "Specification on Steel Rails," the beginning of a large organization for the standardization of all types of materials and tests, including metals, cement, paint, petroleum products, glass, textiles, hardness tests, tension tests, and chemical analytical methods.

The 1924 symposium, recorded in the *Proceedings of the Twenty-Seventh Annual Meeting of ASTM*, is the first record of ASTM involvement with the stainless steel industry. By the year 1924, there were approximately 25 ASTM technical committees. The annual meetings, attended by all committee members, were held in June so that professors who were committee members would be finished with their classes. Meetings were often held at the Chalfont-Haddon Hall in Atlantic City, as was the Great Stainless Steel Symposium.

The 1924 symposium was attended by virtually all of the stainless steel producers in the United States and Canada, as well as a speaker from the Krupp Steel Works in Germany. The symposium also apparently triggered the idea of forming an ASTM committee on stainless steel, an event that took place in 1929. The 1924 symposium was undoubtedly the first large gathering of people in America from the stainless steel industry, an industry that had just gotten started by the Firth-Sterling Steel Company at McKeesport, Pennsylvania, in 1915. Thus far, in America, there was only the production of cutlery steel, the martensitic chromium class of stainless steel. Half a dozen papers on the chromium stainless steels were presented, followed by a paper by Dr. Benno Strauss (Fig. 4), who had come from Germany to describe the iron-chromium-nickel alloy that he and Eduard Maurer had developed.

Data on Stainless Steels Presented

The symposium was introduced by the organizer, Jerome Strauss of the Vanadium Corporation, who also presented the first paper. Mr. Strauss decided that one of the first things that needed to be done was to collect data from as many companies as possible on the properties of alloys being produced for their corrosion, heat, and electrical resistance. There were no reference books with these kinds of data. The first book on stainless steel had yet to be written, except for a trade publication by Firth-Sterling Steel that had come out the previous year, in 1923.

Mr. Strauss had obtained data from 10 stainless steel manufacturers and 22 manufacturers of nonferrous metals and alloys, such as Duriron, Stellite, and Nichrome. Many manufacturers submitted data on several alloys that they produced. The data included the chemical analysis, information on heat treating and working, mechanical prop-

Fig. 4 Benno Strauss, who promoted the industrial application of chromium-nickel austenitic steels that he developed with Eduard Maurer at Krupp laboratories from 1909 to 1912. Source: Thum, 1933, p 374

erties from 0 to 1000 °C, electrical resistance properties, and the corrosion resistance in air, water, and a variety of chemical solutions. Mr. Strauss compiled all of the data and printed it on three large charts that were circulated at the meeting for discussion.

The data on stainless steels were submitted by the following companies: Allegheny Steel Co., Carpenter Steel, Crucible Steel, Firth-Sterling Steel, Ludlum Steel, Midvale Steel, Michigan Steel Casting

Co., and the Canadian company, Atlas Steel Corp. The alloys for each manufacturer were identified by trade names such as Ascoloy, Atlas Stainless, and No. 17 Metal. AISI numbers were not yet developed for stainless steels.

The entries for chemical composition were interesting and showed reluctance on the part of some companies to reveal what they may have considered to be trade secrets. One company listed no chemistry. One company listed only a chromium content, and several listed only chromium and carbon contents. One company listed a broad range for each element, such as "0.18 to 0.70% carbon." Only three companies gave values for carbon, manganese, chromium, and nickel. It was also noted that only one company, Firth-Sterling Steel, listed one of their two alloys as "stainless iron" but did not list the chemistry.

History and Patents

The second paper presented was by P.A.E. Armstrong (Fig. 5), Vice President of Ludlum Steel Co., Watervliet, New York. The Ludlum Steel Co. had been marketing its own stainless steels with the trade names Neva-Stain and Silchrome from the earlier work by Armstrong as a metallurgist with the Ludlum Company. These alloys were high chromium, low carbon, silicon, and iron and had been patented in 1919 by Armstrong.

At the symposium, Armstrong presented an excellent 13-page paper that covered a history of stainless steels and other corrosion-resisting alloys and included many references to patents, both American and foreign. Armstrong mentioned the Krupp patent for "alloys of chromium, nickel and iron, with chromium content of from 1 to 25 percent, nickel from 0.5 up to 20 percent, and carbon from zero up to 1 percent." It is stated in the patent that "these alloys remain bright even when subjected to damp air for a period of months."

Mr. Armstrong failed to realize the significance of the Krupp steel, probably because of the way the patent was written. Benno Strauss clarifies the importance of his steel, as shown subsequently in a discussion of the Krupp alloy.

An Iron-Chromium-Nickel Alloy

The third paper was presented by Dr. Benno Strauss of the Krupp Co., Essen, Germany. It is the only paper on iron-chromium-nickel al-

Fig. 5 P.A.E. Armstrong, who developed silicon-chromium steels used for gas engine exhaust valves. Source: Thum, 1933, p 486

loys and must have been of special interest, because this class of stainless steel had not yet been used in America.

Strauss actually discussed two alloys: V1M, a martensitic chromium alloy, which was a cutlery steel containing a small amount of nickel, and V2A, an austenitic alloy with 15 to 40% chromium, 1.0% maximum carbon, and 4 to 20% nickel (German Patent 304,159). He stated that it was especially useful for resistance to acids. (The alloy that Krupp was actually making was 20% chromium and 7% nickel, similar to the modern type 304 alloy. This information appeared under a micrograph that he displayed.)

Strauss described the V2A alloy and presented data indicating its complete resistance to corrosion in a 30% solution of boiling nitric acid. He also submitted a quarternary (carbon-chromium-iron-nickel) phase diagram that illustrated the required percentages of nickel to create a fully austenitic structure, with varying amounts of chromium up to 24%. He mentioned that "low carbon content increases corrosion resistance." He explained that goods made from his alloys were first shown at the Malmö Exhibition in April 1914.

He concluded in his paper that "austenitic V2A steel is used for acid pumps, valves of all kinds, piping, kitchen utensils, spoons, forks, table utensils, beer casks, Hollander knives, knives for cutting glue, surgical and dental instruments, and metal mirrors." He also noted that:

> "Mirrors and other polished articles when made of V2A steel retain their polish permanently. Owing to its ductility, V2A steel permits of being rolled, forged, stamped and drawn out in the cold state, so that it can be made into thin sheets, wire and wire gauze.
>
> "This brand is well adapted for a great variety of uses in the chemical industries, where it is called upon to resist the attack of nitric acid, ammonia, or hydrogen peroxide in the presence of steam. Its resistance to diluted sulfuric acid is not satisfactory."

The reaction to Strauss' presentation during the symposium is not known, but it must have been well received. Within four years, this austenitic alloy was being made in the United States under the German patent. A modified version of Krupp's 20-7 alloys was then under development in England, an alloy called 18-8 or type 304, which is today the most widely used of all stainless steel alloys.

Continuing Role of ASTM

After the 1924 symposium, ASTM established Committee A-10 on Corrosion-Resistant and Heat-Resistant Stainless Steel in 1929. The first specifications of Committee A-10 were A 167, "Iron-Chromium-Nickel Stainless Steel Plate, Sheet and Strip," and A 176, "Iron-Chromium Stainless Steel Plate, Sheet and Strip." At the time of publication in 1935, the Chrysler Building in New York City had been built with a strikingly beautiful tower clad with stainless steel, the first major use of stainless steel in architecture. Members of Committee A-10 looked on that tower as an ideal outdoor exposure test site for the new metal, because it was exposed to a rather aggressive environment, being close to the sea and the sooty output of a nearby coal-fired steam plant. Five committee members volunteered to serve as a task group to inspect the tower at five-year intervals to determine the extent of corrosion, if any.

In the 1960s, when the American Iron and Steel Institute (AISI) discontinued assigning numbers to new steel alloys, including stainless steel alloys, Committee A-10 filled the gap by assigning an XM series of numbers so that new stainless steel alloys could be included in ASTM specifications. At the urging of ASTM Committee B-2 on Nonferrous Metals, which needed a numbering system, especially for nickel alloys, ASTM and the Society of Automotive Engineers (SAE) eventually developed the Unified Numbering System for Metals. The new system filled the need of the stainless steel committee for a numbering system as well as a general need throughout most of the metals industry.

In 1972, Committee A-10 merged with Committee A-1 on Steel to form a prestigious committee of 900 members. Stainless steel standardization continued to grow, particularly with the acceptance of foreign members and the acceptance of stainless steel alloys from France, Germany, Italy, Japan, Sweden, and the United Kingdom in ASTM specifications. Today, there are over 200 ASTM specifications and test methods for stainless steel and over 150 stainless steel alloys. Stainless steel standardization continues to be one of the most active activities within ASTM.

CHAPTER 5

The Life of Harry Brearley (1871–1948)*

"The reader will observe that my early work on high chromium steels was not inspired by any intention or hope on my part of discovering a stainless steel."
Harry Brearley, Torquay, 1929

BREARLEY'S AUTOBIOGRAPHICAL NOTES were written in 1929, when he was 58 years old and convalescing from an illness at Torquay, a seaside resort in Devon. The papers were bound and sent to his only son, Leo. In 1988, at the time of the 75th anniversary of Brearley's discovery of stainless steel, British Steel Stainless planned a commemoration and began collecting memorabilia concerning Brearley and his work. Brearley's notes were found in Australia in the possession of his grandson, Basil. The notes were published as *Harry Brearley—Stainless Pioneer* by British Steel Stainless and the Kelham Island Industrial Museum in time for the 75th anniversary celebration.

Readers will discover how Brearley's early life and experience led him to become a self-trained chemist and metallurgist. Outokumpu Sheffield (a successor company to British Steel Stainless) has kindly granted permission to quote portions of Brearley's autobiography.

The Early Years

"I was born on February 18, 1871, in a backyard off Spital Street, Sheffield, in what was called a 'House, Chamber and Garret' dwelling.

*Permission granted by Louise Fairweather, Outokumpu-Sheffield (successor company to British Steel Stainless), to use excerpts from *Harry Brearley—Stainless Pioneer*, published by British Steel Stainless, 1988

The house would be less than twelve foot square. How we lived I do not know, and I do not know how my mother managed to keep us reasonably clean.

"I am the eighth child of John and Jane Brearley's nine children and the youngest of five sons. My elder sister had gone into domestic service before I began to take notice. My earliest recollection was of my crying myself awake. After comforting me my mother took me to take back some washing. My mother did other people's washing to help keep the home going, wonderful woman.

"I should be about five years of age when we moved from the back-yard in Spital Street into a front house in Marcus Street. Instead of playing in a crowded court we had the run of a quiet street and plenty of spare land running steeply down to the railroad sidings. This was for some years a wonderful playground, ample, varied and stocked with raw materials for every game we knew or could invent."

Harry's Schooling. "I began to go to school about this time, going with my brothers, but all I can remember of these earliest school days is that we marched towards the school exit at noon singing 'Home to dinner, Home to dinner, There's the bell, There's the bell, Mash a tater in a can, Ding dong bell, Ding dong bell.'"

At a very early age, Harry expressed a keen interest in the working of metals. "I learned many things in those early days as I made my roundabout way home. I spent hours watching pocket blades forged, files ground, lead toys cast and metal buffed and burnished to shining brightness.

"I have no idea of how I learned to read. My father and mother and brothers were readers of novelettes and blood and thunder stories. But there were no books at home, absolutely none. Although I had less than average schooling as a child I never clamored for more. As a child I was delicate. I remember hearing my schoolmistress telling my mother I was delicate. I didn't know what it meant but I realized that it was some disadvantage which might prevent regular attendance at school."

Harry describes the work of his father and grandfather in the metal trades and ironmaking. "My grandfather was a country blacksmith and my father was a steel melter. I suppose I had heard my father and mother talk about smithy work and hardening steel by cooling it off in water.

"Home lessons for the next day's school had no terrors for me. I did no home lessons. I was caned occasionally but that was no penalty for the freedom of the streets which I enjoyed. Neither my father nor my mother ever mentioned home lessons; it would have been an impractical proposal. There really was no room indoors for the family, to say nothing of a vacant table. There was not even a chair apiece. As a child I never thought of sitting on a chair. We youngsters stood round the table at meal times and disappeared out of doors as soon as meals were over, wet or fine, summer or winter.

"Since I 'got on' I find myself constrained to do many things which are not worthwhile, and one of them is to sit in a stiff chair when I would rather squat on the floor.

"My mother had six months of schooling as a girl but she could read and write. Whatever driving force there is in me came from my mother. My father was a dreamer, strong, industrious and able at his job or at anything to which he turned his hand, but entirely lacking in ambition and the considered use of his talent for the support of a large family. We were always poor even when my father had regular work. We never starved in the sense that we wanted bread, but it ran to bread and butter on Sundays only."

Harry's First Jobs. "It was a great time when I was able to add a copper or two to my mother's purse. I began regularly to do so by bundling sticks. A few boys and girls in Marcus Street were allowed to bundle sticks after school hours. We were paid a penny per dozen bundles but we often had to chop our own sticks. The owner of the stick shop was a timber yard labourer who made and sold bundled sticks as an extra.

"Before I was eleven years of age I had passed the 6th standard and was entitled to leave school. First began the job at Marshland's Clog Shop. My job was to do nothing in particular but everything I was asked to. My job was fetch and carry. The day began at eight and finished at eleven. My next job was at Moorwood's Iron Foundry where I was warehouse boy. Having the run of the works was full of interest—put black varnish on kitchen stoves—had to leave when found to be below factory age.

"My next job was as cellar lad in the crucible steel-making furnaces. I was a slim, fair-haired, blue-eyed boy and altogether too frail, the workmen thought, to make a cellar lad, much less a steelmaker. My father never expected me to be a steelmaker, as he believed I was not strong enough. I ceased being a cellar lad because I was below the minimum age regulation of the Factory Act."

Harry Becomes a Bottle Washer. "In October, 1883, when I was just twelve years of age, a newly-appointed chemist took charge of the laboratory at the Norfolk Works of Messrs. Firth & Sons. The new chemist, James Taylor, was a dark, thin, pale man, thirty-five years of age. A few weeks after he arrived he wanted a boy to wash up and I was chosen. The first day I spent with Taylor left an impression I never shall forget. I had never been in a lab before. In fact, I had never heard the word 'laboratory' and had no idea what purpose the place served once I got inside it. There was so much glassware about I thought it might be a room where something to drink was prepared.

"The first days in the lab were unspeakably dreary. Taylor was not the kind of person to arouse interest and confidence in the undisciplined boy. During the second week Taylor asked me what I read and I said 'Boy of England, the Comic Journal and Jack Harkaway.' He appeared not to have heard of these papers. Eventually I worked diligently at whatever book Taylor suggested to me. The first was 'The Irish National Arithmetic.' With this little book I began at simple addition and worked to the end. Taylor was pleased with my industry and I was pleased because he would talk about arithmetic, although he had nothing to say about 'Jack Harkaway.'

"In 1885, when I was fourteen, Taylor bought me a copy of 'Todhunter's Algebra,' a large book of about 600 pages which cost 7/6 d. I was touched that anyone should think it worthwhile to give a book costing so much money. I can see myself proudly taking it home and showing it to my mother who was swilling down the pavement after a load of coal had been delivered. I still have that book.

"Taylor suggested that I should attend night school and study math. Whatever he suggested I attempted, he was my king and could do no wrong or think it either. I attended night school two or three evenings as week and worked at my books until I was driven to bed.

"I was a bottle washer to Taylor and two assistants. It was an easy job with spare hours I could devote to school work which consisted mostly of math and general physics. One of the lab assistants was Colin Moorwood. I owe a great deal to him. He sang snatches of opera in the lab and recited bits of poetry. He smiled when I asked if Shakespeare was an Englishman, but understood sympathetically the depths of my ignorance. He understood the Sheffield dialect but taught me to speak English and I learned from him the music of words and some verses full of high aim and chivalry. Colin was a gent who taught me by example and has remained one of my best friends. He had been brought up in a comfortable home. I knew his home, first as a boy when I would take a message and receive a chunk of sweet cake from his dear mother, and later when, being neither man nor boy, I was learning to feel at home among 'educated' people.

"I learned gradually to make whatever pieces of apparatus might be required. Taylor gave me my first lessons in glass blowing and, in time, I learned to do it better than he. I made a habit, before work in the lab commenced, of working with the joiners and plumbers so that by the time I was twenty, although I had no trade, I was quite at home working with many handicraft tools."

Brearley Becomes a Laboratory Assistant. "I had become an assistant in the lab before Taylor left England and I lost my mother about the same time. My mother's death caused our home to be broken up and I found myself in lodgings with my brother Arthur and, as usual, he was keeping an eye on me. I became interested in the girl who ultimately consented to marry me, and whose feet now adorn the fender on which mine are now resting. I had attended a bible class at the Sunday School. It was one means of meeting Nellie but it had other attractions. Some of the young men were very wide awake. They were interested in talking and a few of them talked of strange books. There was a mutual improvement class on Saturday evenings where good speeches would sometimes be made. I was so much attracted by some of Ruskin's books that I neglected everything else to read them and to read some of the intelligible Carlyle. Ruskin's 'Unto the Last' was a revelation. Ruskin's 'Analytical Economics' and Todhunter's 'Algebra' are the two books that I prize above all others. This excursion into literature excited an appetite which will never be satisfied. But I saw no living in it and I was really equally greedy to understand some of my work from which interests I had temporarily separated myself. There was some prospect of becoming an analyst which I could not afford to neglect."

Brearley Rises to Analytical Chemist. "I decided to understand chemical analysis thoroughly and began reading about the determination of manganese in steel. I attacked one steelmaking interest after another, until at the end of six years I knew my special subject backwards and forwards. I was so poor I had to provide a week's lunches out of 18 p—a brown loaf and dates costing 4 p half penny would serve for two days. It was a regular B. Franklin existence, but my teeth were good."

"At twenty-four I thought about getting married, now earning £2 a week. Had I been asked how I proposed to keep a wife I should have shown my hands and said, after Abernathy, 'with these and a determination to use them.' We were a practical couple with no expensive habits, who were very much in love with each other and delighted to be in the country away from town life.

"After the excitement of marriage I settled down again to reading chemistry. I began to write about the analysis of steelworks materials. Between 1895 and 1902 I contributed scores of articles to *Chemical News* and other journals. Taylor wrote me from Australia in 1897 offering me a position in the assay lab dealing with gold and silver and the lead-antimony-tin alloys generally known as white metals, but I was to study the assay and analysis of the metals before going. I decided to turn it down."

Brearley's First Book. As a young man of 31, Brearley was working as a chemist at Kayser Ellison's steelworks in 1902. "I was their first chemist and Mr. C.W. Kayser kindly allowed me to do some private analytical work. They were in the tool steel business. Whilst there I wrote, with Fred Ibbotson, my first book, *The Analysis of Steelworks Materials*. The actual writing was done in six or eight weeks, but I was so full of the subject and wrote easily, so long as I ignored tenses, moods, split infinitives and other niceties of composition of which I had no real knowledge. Ibbotson thought the English of that part of the book that I wrote was imperfect but he said there was blood in it. This was almost literally true. When I was finished I felt exhausted; I felt as a woman must feel after giving birth to a wanted child—exhausted and triumphantly glad."

About this time, Brearley became involved in a part-time venture with Colin Moorwood, which they called Amalgams. Brearley had discovered how to produce a certain claylike material that a large company would buy from him. "Every evening and every weekend I worked for the Amalgams Company, sometimes in the workshop and sometimes at home. One room of our house was littered with experimental bits and pieces and packages of the material; and then it was improved so much out of recognition as to be a new material. Within two years the business warranted the engagement of a man to look after it; I had other things to do."

Brearley Becomes a Chemist at the Riga Steelworks. "At about this time Firth's bought a partly erected steelworks at Riga to which John Crookston, their Odessa agent and a naval architect by profession, was appointed manager. Crookston and Moorwood met in Riga to consider how the partly built works should be completed and started. On Moorwood's recommendation, Crookston offered me a job as the chemist of the Riga Works. I accepted the job and prepared to leave for Russia in the new year of 1904. I was then thirty-three years of age. About the middle of January Colin and I traveled overland to Russia."

Brearley busied himself setting up an analytical laboratory and buying the various chemicals and pieces of glassware needed. The place was unheated, and because it was in the dead of winter, he worked in an overcoat, a dressing gown, and a long pair of rubber snowshoes. He had a portable, smelly, paraffin stove that he moved from place to place as he worked.

The Russo-Japanese War had been declared by the time Brearley reached Riga, and this war directed the attention of the Salamander Works to the manufacture of armor-piercing projectiles. Billets were to be imported from Sheffield, forged to shape, annealed, machined, and hardened and tempered. The factory was ill equipped and had no experienced men.

Firth's sent out a man from Sheffield named Bowness, who was said to know all about hardening shells, but he proved to be incompetent. His shells failed at the firing tests at St. Petersburg, and Bowness was promptly sent back to Sheffield. Brearley was then offered the job of shell hardening, which he accepted without hesitation. Brearley's brother, Arthur, who was overseeing the building of the melting furnaces, took over the laboratory. When back in Sheffield, Harry had trained his brother in the art of analyzing steel.

Brearley is Promoted to Heat Treater and Then to Works Manager at Riga. Harry needed to test samples of the shell steel that were hardened at different temperatures to find the ideal hardening temperature, but there were no thermocouples or any other pyrometric devices. The temperature of the hardening furnace could only be determined by eye. Harry thought it might be possible to find some salts that would melt at different temperatures. By mixing chemicals in the laboratory, he finally found three metallic salts that melted at approximately what they thought might bracket the hardening temperature. Each of the three mixtures was melted and cast into small cylinders that were then coated with brown, green, or blue wax to identify them and protect them from moisture. They called the little cylinders "Sentinels" or "Sentinel Pyrometers."

Harry sent the formulae for the three Sentinels to his Amalgams Company in Sheffield and said that his little invention resulted in the production and sales of thousands of the Sentinels.

Under Brearley's supervision, and while using the Sentinels, the first batch of shells that was heat treated passed the firing test, and they never again failed to pass the test. After his success with shell hardening, Brearley was promoted to Works Manager. It had been just one year since his arrival in Riga. He spent three more years in Riga, leaving for Sheffield in 1909.

Brearley Becomes Manager of Firth Brown Research Laboratories

In an unusually cooperative spirit, Thomas Firth & Sons and John Brown & Company, two neighboring steelworks in Sheffield, decided to set up a research laboratory that they would operate jointly. When Brearley returned from Riga, he was offered the position of research director. Brearley accepted the job, although he had other options. He still owned the Amalgams Company and had some other ideas about doing private consulting. He was then 38 years of age.

Before accepting the job, however, Brearley thought to have a clause in his contract stating that "any discovery and patents resulting from his work shall be the property jointly of the Company and Brearley."

By now Brearley was a skilled analytical chemist and a self-trained metallurgist, by virtue of his experience as Works Manager of the Salamander Plant at Riga. However, he admitted that he knew least about the physical changes taking place during the heating and cooling of steel. The little he did know related to the influence of heat treatment on the mechanical properties as determined by tension testing and "nicking and breaking." He admitted to being completely ignorant of the microstructure of steel and more than a little confused.

Brearley was given free reign on his research projects and was even given the option of turning down any job in which he was not interested. He set about studying heat treating and became much involved in Izod notched-bar impact testing, ordering two of the earliest machines. He went on to study armor plate. He wrote a book, *Tool Steel*, and another, *The Case Hardening of Steel*, that even had a German edition. He thought his finest work was the book *Ingots and Ingot Moulds* that he co-authored with his brother, Arthur.

In May 1912, Brearley visited the Royal Arms Munitions factory at Enfield to investigate the erosion and fouling of rifle barrels. He had worked with steels containing 5 to 6% chromium and thought steel with more chromium content might be a solution to the rifle problem, particularly because of the high melting point of chromium. At that time, chromium steels were already being used for exhaust valves in airplane engines.

Brearley made a number of different melts of 6 to 15% chromium with varying carbon contents, using first a crucible process and then an electric furnace. The first steels later to be recognized as stainless were melted in August 1913. At this time, Brearley was still trying to find a steel with more wear resistance. However, he then discovered that the new steel strongly resisted chemical attack, after samples of the steel were polished to a mirror surface and etched with a nitric acid solution (as part of the process to examine the crystalline structure of the steel at high magnification).

Brearley Discovers Stainless Steel. On August 20, 1913, Brearley made a cast of steel (No. 1008) having 12.8% chromium, 0.24% carbon, 0.44% manganese, and 0.20% silicon (as noted by Dr. K.C. Barraclough in his account of the 75th anniversary of Brearley's discovery). This was the first commercial cast of what came to be called stainless steel. An ingot was forge cogged to 3 inches square and rolled to a 1½ inch diameter bar. Twelve sample gun barrels were forwarded to the arms factory.

Although Brearley's steel did not prove useful for gun barrels, he described the moment of his actual "discovery" of stainless steel as follows:

> "When microscopic studies of this steel were being made, one of the first noticeable things was that the usual reagent used for etching the polished surface of a microsection would not etch or etched very slowly. I found, moreover, with both the usual reagents, that different pieces would etch and some would not etch. I was satisfied, therefore, so far as corrosion with the etching acids was concerned, that merely to specify the composition of the material was not sufficient, because from the same bar of steel I could cut a piece and then a second piece and a third piece, and in different conditions of heat treatment some of them would etch and some would not. The significance of this is that etching is a form of corrosion, and the specimens behaved in vinegar and other food acids as they behaved with the etching reagents."

Brearley also noticed that cut samples of the steel left in the laboratory did not rust. He described his remarkable discovery 16 years after the event in a rather matter-of-fact manner. He also made the very significant discovery that the chemistry and heating and processing of steel are all equally important.

From the end of 1913, Harry Brearley talked of his idea of producing cutlery. Until that time, cutlery had been made of carbon steel. The town of Sheffield also had been the center of the cutlery industry in England for 300 years. Brearley recognized that his new steel had definite advantages over the 25% nickel steel formerly used to resist tarnishing, because it could be softened for machining and subsequently heat treated. However, the making of cutlery out of his steel was not easily accomplished. Two Sheffield cutlers by the name of George Ibberson and James Dixon, who were originally sent bars from the first cast, reported that it was almost impossible to forge, difficult to harden, and dirty when polished.

There also was definitely a prejudice against the idea of producing cutlery blades that would not rust. One of the foremost cutlers of the city said, "It would be counter to nature." G.E. Wolstenholme, one of Firth's directors, downplayed the whole idea by saying, "Rustlessness is not so great a virtue in cutlery, which, of necessity, must be cleaned after each use."

> "Perhaps the idea of producing on a commercial scale a steel which would not corrode sounds ridiculous, at least my directors failed to grasp the significance of it."
> Harry Brearley, Torquay, 1929

Brearley's directors at Firth's couldn't see any future in stainless steel. They refused Brearley's offer to supply heat treated blanks ready for grinding into knife blades, and they refused to apply for a patent for something they deemed useless. In hindsight, the situation was understandable. The cutlers and the steel manufacturers thought that steel that never rusted would adversely affect the demand for their products. This may be so from their perspective as a cutlery supplier (but not from the customer view, nor with demand from dozens of new products that would also be made of the material).

Brearley was left in a quandary, not knowing which way to turn but convinced that the problems of making cutlery from his material were not insurmountable. In June 1914, Brearley happened to be introduced to Ernest Stuart, who was the cutlery manager at the firm of Messrs. R.E. Moseley in Sheffield. Mr. Stuart was skeptical that a rustless steel could possibly exist, but he agreed to work the small sample provided into a few cheese knives.

"A week later Mr. Stuart produced the knives he had made and pronounced them to be both rustless and stainless." Stuart is said to have been responsible for introducing the term *stainless* when, after testing the steel with vinegar, he said, "This steel stains less." Prior to that, Brearley had referred to his invention as "rustless steel." However, after the first attempt, Stuart also said, in unprintable language, that the steel was very hard and that all his stamping tools were ruined. Brearley said that those first knives were still in use in his home 15 years later and looked as good as new.

Then, Mr. Stuart tried a second time and produced finished knives without damaging the forging and stamping tools. However, the knives were very hard and brittle and were similar to cast iron when fractured. Mr. Stuart had made the steel hot enough so that it would work easily but so hot that it became brittle upon cooling. Brearley himself was invited to attend a third attempt. He admitted that he knew nothing of knifemaking and had not previously witnessed the making of a single table blade. However, he knew the temperature at which this particular steel should be worked and hardened. He assisted in making a dozen blades.

Brearley immediately recognized the practical uses of the new material. In a report dated October 2, 1914, he wrote, "These materials would appear specifically suited for the manufacture of spindles for gas or water meters, pistons and plungers for pumps, ventilators, and valves in gas engines and, perhaps certain forms of cutlery." Thus, the discovery of chromium steel as stainless steel is attributed to Brearley's research. In this same period, others in Germany (Maurer and Strauss at Krupp) and America (Haynes, Armstrong, Becket, and Dantsizen) were also experimenting with alloys of high-chromium steels for possible commercial use.

The steel received quick acceptance by the cutlery industry in the Sheffield area. Mr. Moseley ordered the Firth-produced steel through the Amalgams Company and, within two months, had seven tons on order. He would have liked to have had a monopoly, but by then, other cutlers were lining up to get into the business. "The steel declared dead and nigh worthless was made an absorbing topic of conversation among cutlers and steelmakers." Harry Brearley declared Moseley's to be the first successful producer of what was beginning to be called stainless steel cutlery. In January 1915, a brief announcement of the new metal appeared in the *New York Times*. Firth also promoted the stainless steel in a 1915 advertisement (Fig. 6).

Nonetheless, Brearley had felt wronged by the Firth's directors, who refused to supply blanks and apply for a patent. Because of this situation, Brearley tendered his resignation on December 27, 1915, giving six months' notice, which was accepted. Brearley was 44 years of age. He was not worried about the future, but he was saddened to be separated from the new steel to which he had lost his heart. Above all, he

Fig. 6 Early Firth advertisement (1915). Designed by Evelyn D. Roberts, Pittsfield, New Hampshire.

regretted that circumstances forced him to break with a connection that had lasted for seven years. He said, "I need not trouble to describe the attempts made to reach an agreement since none of them was acceptable to me." It was not a happy New Year.

A Stranger Calls. Early in the new year of 1915, from out of the blue, a total stranger appeared one morning on Brearley's doorstep. He was an elderly, white-haired, well-dressed gentleman, 75 years of age. John Maddox was his name, and he had just come up from London. He said that he had connections in the textile business. However, Maddox wanted to know all about stainless steels and was even more enthusiastic than Brearley about the future of them. Maddox said that he knew America well and had considerable experience with patents. He left Brearley with the understanding that he might go to the trouble and expense of getting a patent in Brearley's name. How Maddox had found Brearley is not known.

Brearley was not predisposed to think about patents at that time and dismissed the incident. Maddox, however, turned out to be a man full of energy and became a good friend. When he failed to get a patent, he came back to Sheffield again and again, urging Brearley to apply for an American patent.

Brearley Applies for Patents. Brearley finally set his mind to it and filed for a U.S. patent on March 29, 1915, and for a Canadian patent on April 21. The American application was denied because, with no British patent, the steel was being made by John Brown & Co., Hadfield's, Sanderson, Vicker's, and other Sheffield firms in addition to Firth's and Brown Bayley's. Brearley immediately solicited the help of Sir Robert Hadfield, Dr. Stead, and R.A. Harbord, all of whom provided written statements of their support of Brearley's new application, which was filed on March 6, 1916, and granted on September 5, 1916, excerpts of which are printed in Fig. 7.

Brearley's Canadian patent, which was filed on April 21, 1915, was slightly different from the American patent. Instead of "Cutlery," the title of the patent was "Malleable Steel." The average mechanical properties of the "typical steel," which is actually the composition of the steel of Brearley's discovery in both patents, are for material oil hardened from 900 °C and tempered at 700 °C. The Canadian patent was granted on August 31, 1915.

It is especially interesting to note that Brearley did not try to patent an alloy per se but rather cutlery. This apparently was to overcome the objection of the patent office, which was that patents for chromium steels had already been applied for.

UNITED STATES PATENT OFFICE

HARRY BREARLEY, OF SHEFFIELD, ENGLAND

CUTLERY

1,197,256 Specification of Letters Patent Patented Sept. 5, 1916
Continuation of application filed March 29, 1915. This application filed March 6, 1916

To all to whom it may concern:

Be it known that I, HARRY BREARLEY, residing at Sheffield, Yorkshire, England, have invented a certain new and useful Improvement in Cutlery, of which the following is a full, clear and exact description.

My invention relates to new and useful improvements in cutlery or other hardened and polished articles of manufacture where non-staining properties are desired and has for its object to provide a tempered steel cutlery blade or other hardened article having a polished surface and composed of an alloy which is practically untarnishable when hardened or hardened and tempered. The alloy is malleable and can be forged, rolled, hardened, tempered and polished under ordinary commercial conditions.

A typical composition of the untarnishable steel blades embodying my invention would be as follows: carbon 0.30 percent, manganese 0.30 percent, chromium 13.0 percent, iron 84.6 percent.

What I claim is:

1. A hardened and polished article of manufacture composed of a ferrous alloy containing between nine percent and sixteen percent of chromium, and carbon in quantity less than seven-tenths percent.
2. A hardened and tempered and polished cutlery blade composed of a ferrous alloy containing between nine percent and sixteen percent of chromium, and carbon in quantity less than seven-tenths percent and not containing any microscopically distinguishable free carbides.
3. A hardened and polished cutlery article composed of a ferrous alloy containing between nine percent and sixteen percent chromium and carbon in quantity less than six-tenths percent.
4. A hardened and polished article of manufacture composed of a ferrous alloy containing approximately carbon 0.30 percent, manganese 0.30 percent and chromium 13.0 percent.

Fig. 7 Text excerpts from Brearley's 1916 patent of a stainless steel

When word of this patent reached the directors of Thomas Firth & Sons, they were astonished and immediately foresaw problems in America for their subsidiary, Firth-Sterling Steel Company at McKeesport, Pennsylvania. Brearley's American patent could interfere with the production of stainless steel, which had already been underway at Firth-Sterling for over a year. After considerable debate, Firth's directors agreed that they should offer to purchase a half-share in the American patent. They were only now agreeing to act in accordance with the terms under which Brearley had accepted the position (i.e.,

"Any new facts relative to the Company's manufactures which shall be discovered by Harry Brearley during the period of engagement and any patents based thereon, shall be the property jointly of the Company and Harry Brearley in equal proportion.").

The Firth-Brearley Stainless Steel Syndicate

Brearley agreed to accept the company's offer to purchase a half-share in his patent only if they agreed to his plan to establish a Firth-Brearley Stainless Steel Syndicate, which would be formed "to foster the world-wide production of stainless steel cutlery." This involved Brearley's renewed association with the directors of Firth's, the people he felt had behaved irresponsibly to him, but it seemed quite obvious that Brearley planned to manage the syndicate. The agreement on the syndicate was reached in July 1917, and it stipulated that henceforth "all knife blades made of Brearley's stainless steel alloy shall be stamped with the following logo:"

FIRTH-BREARLEY STAINLESS

In the meantime, Brearley had started his job as Works Manager of Brown Bayley's Steel Works in Sheffield on July 15, 1915. Brearley was then 44 years of age. He threw himself into the work of the new job as if trying to forget the recent unpleasantness. Brearley was kept busy supervising the modification of manufacturing processes. Because of the war, the company was being pressed to undertake the manufacture of special steels, aero-engine crankshafts, and rifle barrels, none of which had been in Brown Bayley's line of work. Brearley also served on several committees associated with the Ministry of Munitions until the end of the war. He worked at Brown Bayley's Steel Works until 1925.

Stainless Steel Becomes Critical for the War Effort. As the Great War progressed and the Kaiser's bombs began dropping on London and other parts of England in 1915, there was a major effort to build fighter planes for the Royal Air Force. Stainless steel was found to be an ideal material for withstanding the high temperatures of aircraft engine exhaust valves when the chromium was increased from 12.5% for the cutlery alloy to 14%.

In 1916, Firth's, the firm that had belittled the value of stainless steel, immediately began marketing Firth's Aeroplane Steel, or FAS. The production of high-chromium steels for other than defense purposes was prohibited by government decree for the duration of hostilities. By that time, most of the steelworks of Sheffield were producing varieties of stainless steel.

The Firth-Brearley Stainless Steel Syndicate Faces an Unexpected Problem in America. An unexpected problem came up in America when Elwood Haynes (Fig. 8) opposed Harry Brearley's patent. Haynes was the president of the Haynes-Stellite Company in Kokomo, Indiana. Haynes had been experimenting with high-chromium, low-carbon steels as early as 1911, primarily to determine if those alloys might make a less expensive cutting tool material than the cobalt-base Stellite alloys of his company. Haynes had filed for a U.S. patent a little earlier than Brearley's first application of March 29, 1915. When Haynes' application was denied, he appealed and requested an interference order, while agreeing that the discoveries were similar and made independently. In the middle of 1917, the Patent Office granted an interference order to Haynes' patent, which he appealed, and finally granted him Patent 1,299,404 on April 1, 1919 (Fig. 9).

Fig. 8 Elwood Haynes, who was a pioneer American automobile maker and an inventor of a series of complex alloys from 1907 to 1913 when searching for durable spark plug alloys. Courtesy of the Elwood Haynes Museum, Kokomo, Indiana.

UNITED STATES PATENT OFFICE
Elwood Haynes, of Kokomo, Indiana, Assignor to the American Stainless Steel Company, of Pittsburgh, Pennsylvania, a Corporation of Pennsylvania
Wrought-Metal Article
1,299,404 Specification of Letters Patent Apr. 1, 1919
Application filed March 12, 1915

To all whom it may concern:

Be it known that I, Elwood Haynes, a citizen of the United States, residing at Kokomo, in the County of Howard and State of Indiana, have invented certain new and useful Improvements in Wrought-Metal Articles of which the following is a specification.

This invention relates to wrought metal articles; and it comprises wrought metal articles of manufacture of the nature of cutlery and edged tools, such articles having polished surfaces of the general character which is termed noble, in that such surfaces are incorrodible, lustrous and of permanent nature and such articles being composed of a worked down and hard body of an iron-chromium alloy low in carbon and other metals, such alloy being stiff, strong and elastic, able to take a cutting edge and having other properties of tempered metal; such as, for example, an iron-chromium alloy containing not less than 8 percent chromium and, very advantageously, not less than 10 percent, and not more than 50 to 60 percent, the best proportion being between 15 and 25 percent, and containing not to exceed 1 percent of carbon (the amount of carbon being advantageously between 0.1 and 0.5 percent) with the rest of the alloy consisting mainly of iron, there being no substantial amount, (say, not over 4 to 5 percent) of other metals than iron and chromium in said alloy; all as more fully hereinafter set forth and as claimed.

The present alloy is well adapted for making pen points for fountain pens, chisels, table knives, forks, and other purposes where hard metal of high temper is required.

What I claim is:

1. A wrought metal tool having polished surfaces of noble metal and composed of an alloy of iron and chromium, with carbon in amount between 0.1 percent and 1.0 percent, said alloy being malleable at high temperatures, hard, stiff and strong at ordinary temperatures and capable of taking and retaining an edge.

2. A wrought metal tool having polished surfaces of the incorrodible character of polished surfaces of noble metal and comprising an alloy containing from 8 percent to 60 percent of chromium and 40 percent to 92 percent of iron, with carbon in amount between 0.1 percent and 1.0 percent, said alloy being readily malleable and workable and being substantially free of other metals.

3. A wrought metal tool having polished surfaces of the incorrodible character of polished surfaces of noble metal and comprising an alloy of iron and chromium containing from 15 to 25 percent of chromium, and carbon in amount from 0.1 percent to 1.0 percent, and being malleable, ductile and elastic.

4. A wrought metal article having surfaces of the incorrodible character of surfaces of noble metal, and composed of an alloy of iron and chromium, containing carbon in amount of 0.1 percent to 1.0 percent, said alloy being readily malleable and workable at high temperatures, and hard, stiff and strong at ordinary temperatures.

In testimony whereof, I affix my signature in the presence of two subscribing witnesses.

ELWOOD HAYNES

Witnesses:
H. R. Perry
R. Crawford

Fig. 9 Text excerpts from the 1919 stainless steel patent of Elwood Haynes

The American Stainless Steel Company (1918–1936)

In the meantime, the patent problem was resolved. In 1917, one of the first actions of the syndicate was to make plans to establish a patent-holding company in America to manage the licensing of the Brearley patent. The American Stainless Steel Company was established in Pittsburgh in 1918, just three weeks before Haynes was granted the interference order. The Brearley people went to Haynes after he had successfully contested their patent and, as a compromise, offered him a position on the board and a 30% share in the new company.

The new company, with an office in the Oliver Building in Pittsburgh, was incorporated in the state of Pennsylvania with the following ownership: Firth-Brearley Stainless Steel Syndicate (40%), Elwood Haynes (30%), and the balance shared equally among Firth-Sterling Steel, Bethlehem Steel, Carpenter Steel, Crucible Steel, and Midvale Steel. Each party would receive a royalty, according to their share in the company, on all of the stainless steel manufactured by the various steel companies licensed to make steel under the Brearley and Haynes patents. The two patents together, although quite similar, would make a formidable barrier to any company that planned to make steel without paying what would come to be quite high royalties, on the order of 20%.

The first president of the company was James W. Kinnear of Pittsburgh, who was a prominent corporation attorney. He served until his death in 1922 and was succeeded by James C. Neale, also of Pittsburgh. There were ten directors in all, with Elwood Haynes having two seats, one of which he assigned to his son, March Haynes.

The first manager of the company appears to have been W.H. Marble. He would be involved in licensing agreements, the collection of royalties, the arrangements for meetings of the board, and the promotion of stainless steel. With regard to the latter, it should be noted that, despite its attractive qualities, stainless steel was not easy to sell in the beginning, and it cost four to five times that of carbon steel.

Early in 1920, Marble gave a paper at the Philadelphia convention of the American Society for Steel Treating. His paper was entitled "Stainless Steel—Its Treatment, Properties and Applications" (*Transactions of the American Society for Steel Treating*, Vol 1, Cleveland, Ohio, 1920). The paper was undoubtedly the most comprehensive yet given on cutlery steel. It contained summaries of recent information from five sources, including Dr. W.H. Hatfield, who was Brearley's successor at Brown Firth Research Laboratories, Elwood Haynes, the

New York Testing Laboratories, the National Bureau of Standards, and the Joint Iron and Steel Committee of the Society of Automotive Engineers and the American Society for Testing Materials.

Marble started out on a favorable note by saying, "During the late war period in England all stainless steel supplies were appropriated by the Ministry of Munitions." He reviewed the chemistry specifications, the effects of various elements, manufacturing, forging, rolling, annealing, hardening, and tempering. He also showed an interesting table of temper colors for stainless and tool steels, which revealed that, for a given color, the stainless steel is always a higher temperature than for tool steel, with the difference increasing for higher temperatures. A greenish-blue color, for example, is produced at 1300 °F for stainless steel and at only 625 °F for tool steel.

A chart was also displayed on scaling tests, which showed weight losses for six alloys when tested in air at temperatures from 1300 to 1560 °F when weighed at 24 hour intervals. Stainless steel showed practically no weight loss after 268 hours at 1540 °F and was lower than either 25% nickel steel or tool steel and much lower than for carbon steel and 5% nickel steel.

A table showed the mechanical properties of stainless steel samples that were hardened from 1650 °F. Five samples each were quenched in air, oil, or water and tempered at five temperatures. The elastic limit, tensile strength, percent elongation, and reduction of area were recorded for each specimen. Although not stated, the data are recognized to be from Brearley's test report of October 2, 1913, on an early cast of stainless steel.

Marble ended his paper with the following conclusion:

> "The steel is costly to produce in the form of cutlery by reason of the necessity for better grinding and more careful forging in consequence to its great hardness. For general use, it may be said that it is costly to manufacture by reason of the expensive alloy employed, and the comparatively heavy waste (due to scrap) involved as a result of very careful inspection; but there *is* a very wide field of usefulness before it."

In 1918, licensees of the American Stainless Steel company included Bethlehem Steel, Carpenter Steel, Washington Steel and Ordnance, and Crucible Steel.

American Stainless Steel Company versus Ludlum Steel Company. Ludlum Steel of Watervliet, New York, was not among the licensees, and when Ludlum Steel began to market its own Neva-Stain and Silchrome stainless steels, there was a possible basis for a lawsuit. These alloys were chromium-silicon steels developed by P.A.E. Armstrong and patented in 1919. The Armstrong alloy was an aircraft valve steel that was sold under the Ludlum Steel trade name Silchrome and later identified in Society of Automotive Engineers' specifications as HNV-3.

In approximately July 1920, F.A. Bigelow, president of Carpenter Steel and one of the directors of the American Stainless Steel Company, sent Elwood Haynes the analysis of the new Ludlum alloy for his comments. Haynes immediately replied, saying it was clearly an infringement of his patent because "the alloy contains more than 8% chromium and was practically stainless." He said his patent was based on "the combination of iron and chromium, and was independent of all other metals, whether added in inappreciable or larger quantities. Neither does it cover the methods of working or heat treating same. Any article made of a combination, therefore, which involves a steel containing more than 8% chromium is an infringement against the patent." The Ludlum alloy was 8% chromium and 3% silicon.

The American Stainless Steel Company board considered the matter and decided to attempt a settlement, but the Ludlum Company stood firm and announced that they would defend any suit based on the use of the unlicensed Neva-Stain cutlery steels. On December 31, 1920, a patent attorney by the name of Christy filed the suit "The American Stainless Steel Company vs. Ludlum Steel Company." The American Stainless Steel Company retained the services of Prof. William Campbell of Columbia College to make some experiments. Harry Brearley also was asked to come to New York, making it clear that just his travel expenses would be paid.

The case came to trial early in1922, with the noted jurist Learned Hand on the bench. Haynes and Brearley testified at different times. Brearley wrote the following in his autobiography about his part in the testimony:

> "When the case opened I was in the witness box for three hours and enjoyed every minute. I had nothing to conceal from the cross-examining and understood the subject and the value of his direct evidence better than he did."

The judge dismissed the case, declaring the patents valid but not infringed by the Armstrong alloys. The Hand decision, which was delivered on April 13, 1922, was reported in *Iron Age*:

> "The issue . . . is whether the addition of silicon, which obviates the additional heating of the plaintiff's composition beyond its critical point, makes the resulting article an infringement. Obviously this could not be . . . Granting that the addition of silicon would not avoid infringement, it does not create."

Haynes felt that the American Stainless Steel Company had proceeded on the wrong basis and listed points he thought should be brought up during an appeal, stressing the idea that neither his nor Brearley's patent covered "any process for either polishing or tempering the alloys." Haynes repeated a point he had presented to his patent attorney: that the primary invention was his "discovery that stainless steel articles can be made from an alloy consisting essentially of 8% or more of chromium and 92% or less of iron." Carbon was not an essential ingredient, but with 0.6% carbon or more, such articles may be tempered or hardened by slightly modifying the manufacturing process. With a chromium content of 10 to 18% and the carbon at 0.6%, "the stainless quality of the article may be enhanced by a suitable heat treatment. The latter feature is covered by Mr. Brearley's patent only, but he gives no specific method for hardening the article."

In October 1922, the Board of Directors decided to appeal Hand's decision and also to dissolve the company if the appeal was lost. On April 16, 1923, the Circuit Court of Appeals handed down a decision that was most welcome to the Board of the American Stainless Steel Company: "Haynes and Brearley both held pioneer patents valid and infringed." In the following statements, Judge Hough gave Haynes credit for first discovering the basic properties of stainless steel:

> "The object of both patents is the same and may be shortly described as a desire to produce what for some years have been increasingly called 'stainless steels.' Although Brearley's patent date is earlier, his date of application is later, and it may be summarily held that Haynes is the generic and Brearley is the specific patent."

Judge Hough gave the reason for reversing the lower court's decision. He reasoned that the defendant has by "omitting silicon . . . produced 'stainless steel;' with the silicon added, he has also produced 'stainless steel;' therefore, in respect of infringement, the silicon is immaterial no matter how beneficial it may be." (Note: A close reading of Judge Hough's reason for reversing the lower court's decision shows that he had come to the erroneous conclusion that all of the alloys covered by the Brearley and Haynes patents, that is, from 8 to 16% chromium, would be stainless steels. However, any alloys today having less than 10.5% chromium would not be considered to be stainless steels.)

Haynes died on April 13, 1925, at the age of 67. The American Stainless Steel Company continued to have good earnings from 1925 through 1929. However, in the early 1930s, there was a suit with the Rustless Iron Company of Baltimore in which the American Stainless Steel Company failed to have their patents upheld. This may have led to their demise, the date of which is not known. In any event, the company would have been dissolved in 1936 when Harry Brearley's patent expired.

Brearley's Later Years

From 1918 on, after the settlement of the Brearley and Haynes patent dispute, the Firth-Brearley Syndicate continued to play an active role. There was the patent infringement suit of the American Stainless Steel Company against the Ludlum Steel Company, in which the Brearley and Haynes patents were involved. Harry Brearley gave some of the most important testimony, but the court decision, at first, was against the American Stainless Steel company. After an appeal, the decision was reversed.

Because Brearley was in America for the trial, he spent some time on syndicate work. He visited a number of cutlery manufacturers and the Firth-Sterling Steel Company at McKeesport, where he had a good discussion with Gerald Firth, the president. Brearley also reported that he assisted them in making a melt of stainless iron. He concluded his visit in America at the Massachusetts Institute of Technology, where he was invited to give a lecture on stainless steel to the students and some interested men from the area.

Brearley also made an extended trip to South Africa, where he gave lectures on stainless steel at Capetown, Durban, and Pretoria. He delivered six lectures at Johannesburg, which were attended by groups from various scientific societies.

A Fraud-Detection Scheme. After Brearley's retirement, the Firth-Brearley Syndicate stayed in business for a while. The syndicate became involved in one of the most unusual cases in the history of metallurgy when they had reason to believe that A.W. Gamage, a London store, was illegally selling knives with blades stamped "FIRTH STAINLESS," which Firth believed were not made of Firth Steel.

Ordinarily, a case of piracy such as this would be difficult and costly to prove, but not so for Firth's, who had foreseen that such an event might occur and had taken certain precautions.

Firth's had been doping their stainless steel with a small amount of element "X," which was added from an unmarked brown packet to each melt by the melting shop manager toward the end of a melt. The chief chemist and one laboratory assistant were the only ones who knew the identity of element "X," nor was its purpose explained if anyone noticed it. Suspected cases of piracy were proved or disproved by one analyst, working alone after hours, who used a color test to detect element "X." Years later, the Brown Firth laboratories revealed that element "X" was cobalt in an amount equal to approximately 0.03%.

Brearley Receives Metallurgy's Highest Recognition. In 1920, the Council of the Iron and Steel Institute presented Harry Brearley with the Bessemer Gold Medal, which is awarded for outstanding services to the steel industry. He was the fourth recipient of the medal. In his autobiography, Brearley wrote, "This is the only distinction I had ever audibly coveted. I valued this presentation all the more because it was made by Dr. Stead, that dear old man whose simple character and manner and life were as admirable as his metallurgical investigations were excellent."

The following is the last paragraph of his acceptance speech:

> "Dependence on the willing help of others is the part of all who successfully direct a laboratory for industrial research. Most problems relating to iron and steel cannot be definitely stated, and the individual, best qualified by experience to study the problem, and to solve it, might be some workman who is engaged on the job day after day. It is the investigator's greatest achievement to inspire interest in such men to make them confederates to his plan. The lust to work, the desire to find out and to understand things is not confined to those who regularly wear a clean collar and, in thanking the Council and Members of the Iron & Steel Institute for this high honour, I am proud to confess my life-long indebtedness to scores of friends, with hard hands and black faces, who toil at laborious tasks in mills and forges."

After the war, Brearley gave up his job as Works Manager at Brown Bayley's but continued to be responsible for the melting plant, heat treatment, and the laboratory.

Brearley Retires. In 1925, although only 54 years of age, Brearley decided to retire. Since returning from Russia in 1907, he had almost always earned more than enough to supply his wants. He had lived simply and never had a fine house, motor cars, or luxurious food. In addition to savings, he still had income from the American Stainless Steel Company in which he held a 40% interest, and he still had income from the Amalgams Company.

He wanted to live more out of doors, play games, and, in general, do things he had never been able to do in his younger days because of lack of time or money.

Harry Brearley retired from Brown Bayley's on the best of terms, so it seems. His salary stopped, but he still came to work and tended to certain company business. However, as time went on, he became more involved with other matters. He assisted J.H.G. Monypenny, chief of the Research Laboratory at Brown Bayley's, in writing the world's first book in English on stainless steel, *Stainless Iron and Steel*, which was published in 1926. Monypenny acknowledged Brearley "particularly for much help and advice."

Within a year or so, Brearley agreed to do some private metallurgical consulting for half a dozen manufacturers who were not steel producers. Brown Bayley's built him a new 30 square foot office, well lighted, heated, and decorated according to his wishes. He was welcome to entertain his clients there, and he also had use of the laboratory. He was soon busier than ever but relished his work with his new clients. Never had a retiree been treated in such an elegant manner by an employer.

In 1927, approximately ten years after the Firth-Brearley Stainless Steel Syndicate was established with the agreement that all blades made with Firth's stainless steel would be stamped with the logo FIRTH-BREARLEY STAINLESS, Firth's dropped Brearley's name from the logo without consulting him. Brearley was furious, but Firth's was adamant. He never completely got over it. The fact that some degree of cordiality had been restored, however, is evidenced by the later writing of Dr. W.H. Hatfield, his successor at Brown Firth Laboratories, whom he must have met, at least on syndicate business. Hatfield wrote:

> "The city has, of course, been famous for its knives for hundreds of years, but down to a generation ago would rust or stain, and needed much attention in cleaning. Nearly all knives now are

made of stainless thanks to the researches of my friend, Harry Brearley, and those who have assisted in the development. This is a delightful instance where the product of a major industry has been revolutionarily transformed by local effort."

In 1929, while Brearley was recuperating from an operation at Torquay on the coast of Devon, he wrote but did not publish his autobiography. He had the manuscript typewritten and sent it to his only son, Leo, who eventually passed it on to his son, Basil. When planning a celebration to commemorate the 75th anniversary of Brearley's accomplishment, the staff of British Steel Stainless searched for memorabilia and found Basil, who was in Australia, and retrieved the manuscript. British Steel Stainless published the book, which was called *Stainless Steel Pioneer—The Life of Harry Brearley*, along with contemporary photographs, which were furnished by the Kelham Island Industrial Museum.

Brearley set up a fund to reward Sheffield authors of creative papers on metallurgy. The biennial prize was to be awarded by the Sheffield Metallurgical Association. He called it the James Taylor Prize, in honor of his first boss.

Another award of recognition came to Brearley in June 1939 when he was 68 years of age. It was the Freedom of Sheffield Scroll and a Freedom of Sheffield Casket, which was a small, ornate metal box that was adorned with six figures engaged in various metals trades.

In 1941, Brearley wrote another autobiography entitled *Knotted String* and published a record of the old Sheffield steel trade called *Steelmakers*. Brearley continued on as a director on Brown Bayley's Board, a position he held until his death in 1948 at age 76.

75th Anniversary. In 1988, the 75th anniversary of Brearley's discovery of stainless steel was organized by British Steel Stainless to celebrate his achievement. As part of the celebration, a scholarly paper entitled "Sheffield and the Development of Stainless Steel" was presented at the Brearley Centre, British Steel Stainless, on October 25, 1988, by Dr. K.C. Barraclough, who had worked on the development of stainless steel while he was at the Firth Brown Research Laboratories.

CHAPTER 6

The Early Books and Papers on Stainless Steel (1917–1949)

IT IS INTERESTING to examine the early literature on stainless steel to gain an understanding of the state-of-the-art of stainless steel as it progressed through the first 37 years after the discoveries made in Sheffield and Essen.

Paper by Dr. W.H. Hatfield (1917)

Dr. William Herbert Hatfield (1882–1943) received his Doctor of Metallurgy degree from University College, Sheffield, in 1913. He is not to be confused with Sir Robert Abbott Hadfield (1858–1940), another famous metallurgist at Sheffield, who is perhaps best known for his invention of Hadfield's manganese steel.

Dr. William Hatfield was the Director of the Firth Brown Laboratories in 1916, succeeding Harry Brearley and continuing Brearley's work on stainless steel.

Hatfield published a report entitled "Heat Treatment of Aircraft Steels" in the *British Journal of Automobile Engineers*, Volume 7, in 1917. This paper showed the results of various heat treatments on the tensile, impact, and hardness properties of a steel containing approximately 0.30% carbon and 13% chromium. The tests included specimens, all of which were quenched from 850 °C followed by air hardening, oil quenching, or water quenching and then tempering at four different temperatures.

The data presented were actually the unpublished data of Brearley recorded in a research report dated December 22, 1913. The data were of particular interest in 1917, during World War I, because Brearley's cutlery steel was being exclusively used for airplane exhaust valves and sold under the brand name Firth's Aeroplane Steel. The heat treatment was critical for this application as well as for cutlery.

Paper on Stellite and Stainless Steel by Elwood Haynes (1920)

Elwood Haynes (1857–1925) of Kokomo, Indiana, is best known for his discovery of the cobalt-chromium-tungsten alloy known as Stellite and his formation of the Haynes Stellite Company, but he also discovered the cutlery type of stainless steel a little earlier than did Harry Brearley. Since his youth, Haynes had had an interest in metals. In fact, his thesis at the Worcester Free Institute in Worcester, Massachusetts, was on the effect of tungsten on steel. His degree was in chemistry, and, like Harry Brearley, his metallurgy was self-taught.

Before Haynes embarked on his metallurgical career, he was a high school teacher for a while, then a manager at the Kokomo Gas Company, and the inventor of one of the earliest gasoline-powered automobiles, which may be seen at the Smithsonian Institution. He cofounded the Haynes Apperson Automobile Company in 1903 and, in 1912, founded the Haynes Stellite Company, which was later bought by Union Carbide and then Cabot Company.

Because of his work on Stellite in 1910 when searching for a better alloy for electric contacts in the automobiles he was building, Haynes became interested in corrosion properties, hardness, elasticity, and cutting qualities.

In 1920, Haynes presented a paper before the Engineer's Society of Western Pennsylvania that was entitled "Stellite and Stainless Steel." The published proceedings consisted of a seven-page paper, with eight pages of discussion, mostly pertaining to Stellite. This was an unusual paper in that two pages were actually quoted from Haynes' laboratory notebook. For example:

> "Nov. 15, 1911. While I have known for some time that chromium, when added to steel or iron, influences or modifies their properties to a marked degree, I am now engaged in gaining a definite knowledge of the effect of chromium on the following:

- The effect of chromium on the resistance of steel and iron to chemical and atmospheric influences
- The effect of chromium on the hardness of iron and steel
- The effect of chromium on the elasticity of iron and steel
- The effect of chromium and the cutting qualities

"The preliminary experiments I have already made along this line indicate that the effect of chromium on iron or steel is much the same as on copper and nickel.

"I have already prepared the followed alloys:

- Alloy 20 C, 79.4% Fe, 20% Cr, 0.6% C
- Alloy 15C, 84.4% Fe, 15% Cr
- Alloy 5C, 95.0% Fe, 5% Cr
- Alloy 10C, 90% Fe, 10% Cr
- Alloy 15C, 85% Fe, 15% Cr
- Alloy 20C, 80% Fe, 20% Cr"

In the following year, Haynes also conducted experiments with four additional alloys, containing 15 to 17% chromium, and reported the results, including in one instance the resistance to nitric acid. Haynes went on to describe the events concerning his immediate patent application in which he claimed that "immune chrome steels must contain more than 8% chromium, though for some purposes they may contain more than that amount, even up to 60%."

Haynes was not actually interested in going into the stainless steel business, because he was totally immersed in his fantastically successful Stellite business. He also had an automobile factory. He had pursued his stainless steel patent application mainly as a matter of pride. In the end, Haynes and Brearley settled their dispute by turning over their patents to the American Stainless Steel Company, a new company formed to manage patents and licenses. Brearley and Haynes were the principal owners of the company.

Stainless Steel Paper by Marble (1920)

In 1920, W.H. Marble, Manager of the American Stainless Steel Company, presented the paper "Stainless Steel—Its Treatment, Properties and Applications" at the Philadelphia convention of the American Society for Steel Treating (*Transactions of the American Society for Steel Treating*, Volume 1, Dec. 1920, p 170–179). The paper begins with a very brief historical account that describes Elwood Haynes and Harry Brearley: "Two men working independently, with no knowledge of what the other was doing, actually developed this unique alloy." He mentions the 1911 work of Haynes and then Brearley's work, although Marble incorrectly attributes Brearley's work to "armor-piercing projectiles for the Russian Government" and to an investigation of "powder-erosion." This was an inaccurate description of Brearley's work.

Nonetheless, the paper by Marble did provide a good summary of previous investigations along with detailed information on the treatment and properties of steel with chromium contents of 11.4 to 14% and carbon contents of 0.20 to 0.40%. As noted in Chapter 5, "The Life of Harry Brearley (1871–1948)," it gave information from five sources, including Dr. W.H. Hatfield, who was Brearley's successor at Brown Firth Research Laboratories, Elwood Haynes, the New York Testing Laboratories, the National Bureau of Standards, and the Joint Iron and Steel Committee of the Society of Automotive Engineers and the American Society for Testing Materials. With the exception of the introductory text on history, the full text of the Marble paper is in the appendix of this chapter.

Firth-Sterling Steel Company Trade Publication (1923)

In 1923, the world's first book, in English, on stainless steel was published. It was a trade publication of the Firth-Sterling Steel Company of McKeesport, Pennsylvania, entitled *Firth-Sterling "S-Less" Stainless Steel.* An earlier book, in French, had been published by Léon Guillet (*Special Steels*, Dunod, Paris, 1905) covering his early research.

Firth-Sterling, a tool steel manufacturer, produced the first commercial stainless steel in America, a heat of cutlery-type stainless steel, in 1915. Firth-Sterling was a subsidiary of Thomas Firth & Sons, Ltd., of Sheffield. It was said that the information in the book came from Thomas Firth & Sons, customers, and their own metallurgists.

It was a hard-bound, 65-page book in a green cover with a gold-imprinted title. It was a remarkable book for its day and was the only book on stainless steel published in America until Ernest Thum's book ten years later in 1933. The book covered Harry Brearley's cutlery type of steel and the nonhardening stainless irons. The specific chemical compositions were not stated, and it was explained that the steels were sold based on their properties rather than their chemical compositions.

The contents included a brief history, applications, chemical analysis, physical properties, and the resistance to 40 agencies of rust, stain, and corrosion. Detailed information was given on forging temperatures and properties after various heat treatments. The properties included tensile strength, yield point, elongation, reduction of area, Brinell and Scleroscope hardness, coefficient of expansion, conductivity, and resistivity. It was a fine handbook for metallurgists and heat treaters of the day.

It was well illustrated, with 14 pages of photographs of applications and several pages of charts. Cutlery (Fig. 10) was illustrated and noted as the first application to take considerable tonnage of stainless steel at the time. The book also illustrated other emerging applications, such as medical instruments (Fig. 11) and golf clubs (Fig. 12). A few of the applications were:

- Airplane parts
- Air compressor valve
- Automobile parts and fittings
- Artificial limb fittings
- Balances
- Ball bearings
- Bicycle parts
- Bits and so forth for horses
- Bolts, nuts
- Bottling machinery
- Bread-making machinery
- Builders' tools
- Butchers' rails
- Carburetor needles, screws, and so forth
- Cooking utensils
- Dairy apparatus
- Dies
- Electrical appliances
- Engines
- Fishing tackle

The book also had a photograph of stainless steel tableware, although the photograph was only of stainless steel knives and forks. The exclusion of spoons clearly indicated that spoons could not be

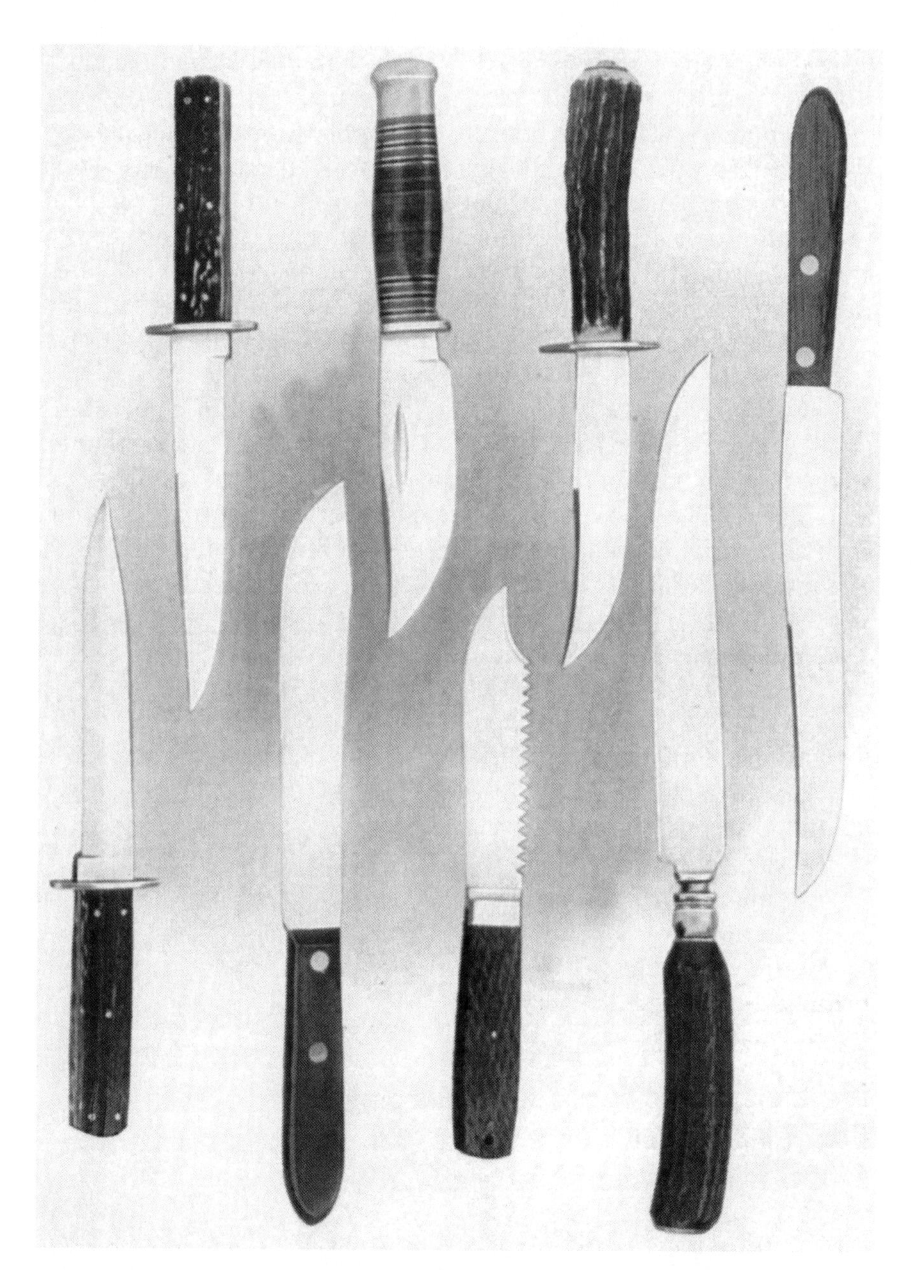

Fig. 10 Stainless steel cutlery illustrated in the book *Firth-Sterling "S-Less" Stainless Steel*, published by Firth-Sterling Steel Co., McKeesport, Pennsylvania, 1923

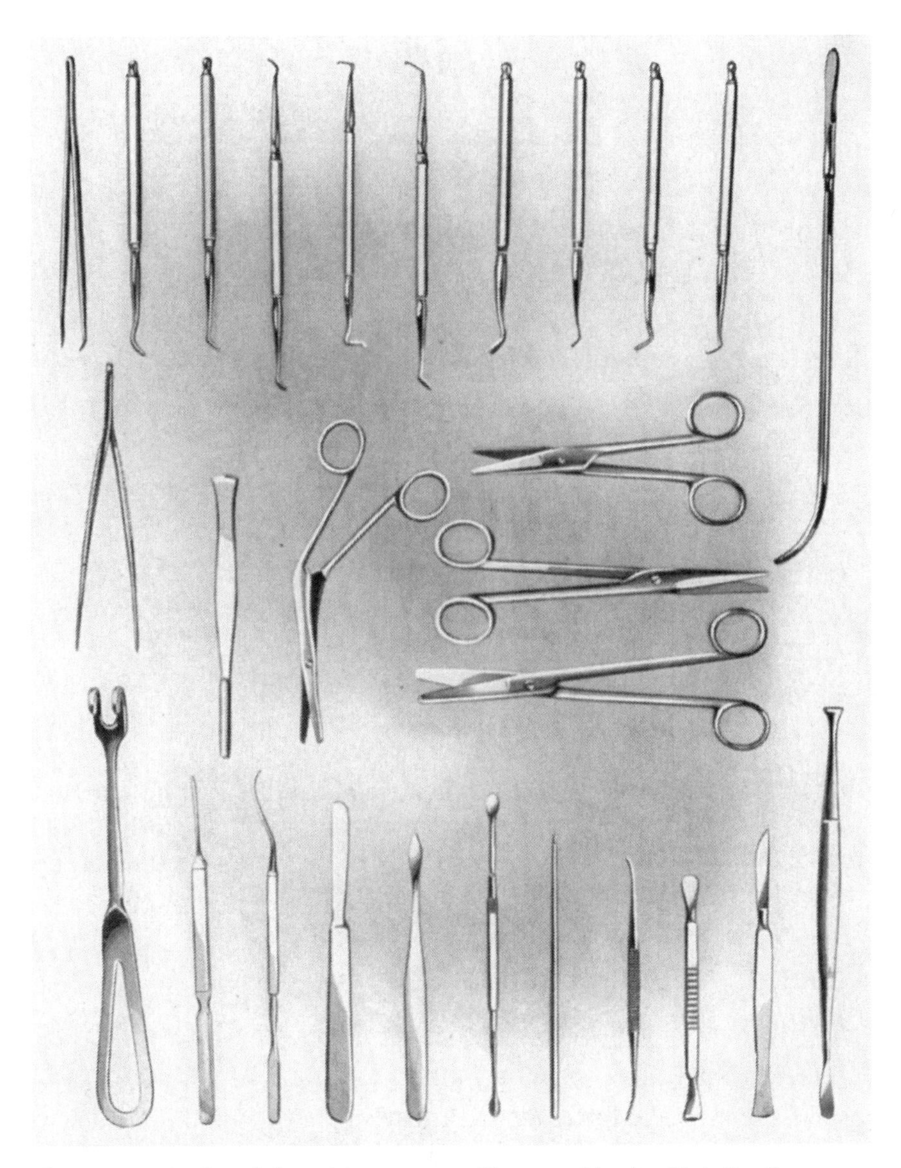

Fig. 11 Surgical and dental instruments illustrated in the Firth-Sterling book, 1923

Fig. 12 Golf clubs illustrated in the Firth-Sterling book, 1923

formed from the cutlery steel. All items had handles that had been attached. The handles of some appeared to be ivory and others were metal. Stainless steel spoons apparently came later with the chromium-nickel steels with the more formable microstructure of austenite.

Paper on Cutlery Stainless Steel by Owen K. Parmiter (1924)

Owen K. Parmiter, a metallurgical engineer for the Firth-Sterling Steel Company at McKeesport, Pennsylvania, delivered an extensive paper on cutlery stainless steels at the Boston Convention of the American Society for Steel Treating in 1924. The paper (published in *Transactions of the American Society for Steel Treating*, Volume VI, p 315–340) is addressed primarily to manufacturers who would be using the steel that his company had been producing since 1915. Little is said concerning its actual production, because the art of steelmaking was a close-kept secret.

A Brief History on the First Decade of Stainless Steel. The Parmiter paper begins by reviewing some of the history in the early development of stainless steel, defined as steel with approximately 13% chromium. It is informative to recount his historical review here:

> "The advent of stainless steel has marked an epoch in the manufacture of tool steel, secondary only in importance to the introduction of high speed steel at the beginning of the present century. Although its development was somewhat retarded during the late war, its worth has now been fully proven under actual service conditions and it has established itself as a valuable commercial steel.
>
> "The knowledge of the value of high percentages of chromium in imparting special properties to iron and steel has been known for many years. As early as 1872, the effect of chromium in producing non-corrosive properties in steel was described, vaguely, however, by two Englishmen, Woods and Clarke, in a provisional petition for a patent which was never granted. Their specification dealt with a low carbon steel, containing approximately 32.00 per cent chromium. Investigation as to the resistance which high chromium steels have for certain corrosive explosives used in ordnance work, was made by the Krupps at their factory in Essen, Germany, as early as 1895. Doctor Goldschmidt of thermit-fame, understood the resistance of such steels to heat and corrosion as well as the importance of low carbon. In an address made at Duesseldorf soon after the close of the Paris Exposition in 1900, Dr. Goldschmidt discussed the difficulties in manufacture, together with the properties and possibilities of the steel.
>
> "In 1900, an exhibit was made by Jacob Holtzer & Company at Paris on a series of steels, containing 0.40 per cent carbon with chromium running from 10.00 to 15.00 per cent. Complete metallurgical data accompanied this exhibit. About this time, a controversy concerning the quality of the alloy required for making this steel arose between the English firm of George G. Blackwell & Sons Company, and the German manufacturers, Goldschmidt Thermit Company. The merits of the two alloys, made by different processes for the same purpose, were ably discussed in the leading English, French and German industrial and metallurgical journals of the day.
>
> "During all this time, little use had been made of the known properties of the alloy, and although it was quite valuable, no important application had yet been made. Elwood Haynes in America and Harry Brearley in England were the first investigators to

successfully apply the useful properties of this alloy steel to every day use. Working independently of each other, they made their discoveries at practically the same time. Both were granted patents on their claims in the United States Patent Office, the application of Haynes pre-dating that of Brearley by only a few weeks. Both applications were filed in March, 1915.

"In the Haynes patent, the composition limits for carbon are stated as being between 0.10 and 1.00 per cent, with chromium from 8.00 to 60.00 per cent.

"The composition limits set forth in Brearley's patent called for carbon under 0.70 per cent with chromium between 9.00 and 16.00 per cent. As a standard composition 0.30 per cent carbon; 0.30 per cent manganese; and 13.00 per cent chromium, was recommended. In justice to Brearley, we can say that in the past ten years this composition has not been improved upon and that it is yet the best balanced stain resisting steel for general every day purposes, although numerous attempts have been made to improve its quality by the addition of various elements or the slight alteration of those already present. Brearley also specifically stated that the steel should be made in the electric furnace, that it should be hardened and polished to give maximum results and that it was specially adapted to cutlery purposes. The truth of these statements has been proven many times since they were first written and today they stand out boldly with almost as much importance attached as the composition itself.

"Although the Brearley and Haynes patents are the most important ones in the field, several others have been granted by the United States Patent Office. In the majority of cases with these latter patents silicon and nickel have been recommended as additional alloying elements. Some patentees advise the incorporation of carbon up to as high as 2.00 per cent. For certain purposes, several of these steels have some advantages, but for all uses where a balance between hardness and stain-resistance is to be maintained, no steel has yet been produced which shows distinct advantages over the standard 13.00 per cent chromium, stainless steel.

"During the period of the war, difficulty was experienced in obtaining chromium due to the fact that the Government commandeered all of the material for war purposes. Little progress in the development of stainless steel was made during that time. Soon after the close of the war, however, rapid progress was made in perfecting methods of manufacture."

Manufacturing Aspects. Parmiter's 1924 paper then continued with a review of some problems involved in the manufacture of this material, its composition, and the effect of various alloying elements. In 1924, he wrote, "Of all the steels known at the present time, stainless steel is possibly the most difficult to produce satisfactorily. Not only is the manufacturer confronted by unusual and extremely narrow compositions limitations, together with a peculiar combination of 'air-hardening' and 'red-hardness' properties, which make working extremely difficult, but experience has shown stainless steel to be most susceptible to surface seams, internal segregates, and non-metallic inclusions of any steel known to modern tool steel metallurgy."

Parmiter then described manufacturing aspects with discussions on composition effects, heat treatment, forging, brazing, welding, and cold rolling. He provided data on mechanical properties, corrosion resistance, and physical properties such as electrical conductivity, magnetic properties, and thermal conductivity. He also preferably described the microstructure of annealed stainless steel as "a uniformly sorbitic structure, although sometimes crystals of complex chromiferrous ferrite segregate themselves." No micrographs were supplied in this 1924 paper.

In conclusion, Parmiter noted that the application of stainless steel was still in its infancy, with many new possibilities beyond their successful use in cutlery. New applications noted by Parmiter were very similar to those described in the 1923 Firth-Sterling trade publication, such as surgical instruments and a wide variety of culinary articles. Parmiter also noted applications such as turbine blading (Fig. 13), high-grade hardware, and machine parts for marine and mining equipment. The industrial significance was very evident at that time, given the earlier use of stainless steel for exhaust valves of airplane engines during World War I.

A Decade Later. Almost one decade later, another paper by Parmiter on cutlery stainless steel, based on an adaptation of his 1924 paper, was published in the first edition of *The Book of Stainless Steels* (E.E. Thum, The American Society for Steel Treating, 1933, p 241). It was also published in the second edition (E.E. Thum, American Society for Metals, 1935, p 278). As in the 1924 paper, he again mentioned that "the cutlery types are probably the most difficult to produce satisfactorily."

Similar to his earlier 1924 paper, Parmiter also began by writing that since Harry Brearley had discovered the composition (0.3% carbon, 13% chromium) for stainless steel cutlery, it had not been im-

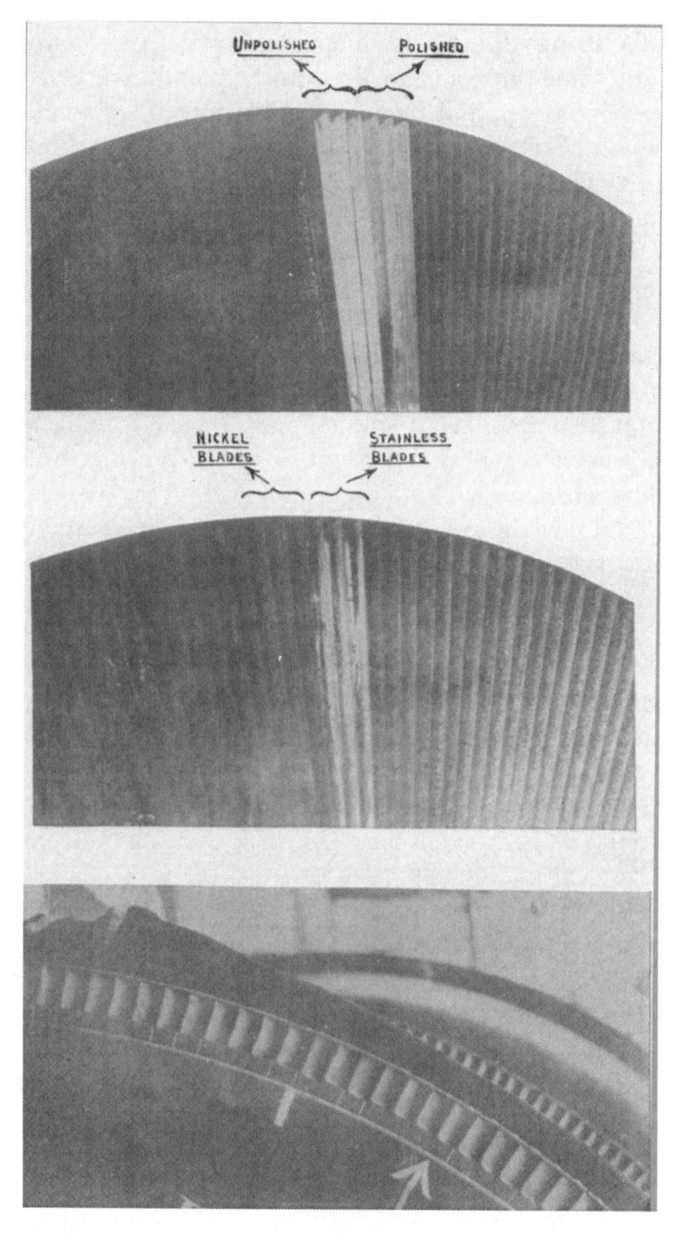

Fig. 13 Stainless steel turbine blading illustrated in Firth-Sterling book, 1923

Effect of Hardening Temperature on Stain Resistance

Oil Quenched From	Brinell Hardness	Resistance to Stain 10% $CuSO_4$ 6 Min.	33% HNO_3 2 Min.	5% Acetic Acid, Dried On
1500° F.	340	50	90	50
1600	440	85	90	50
1700	485	90	90	85
1800	550	95	95	90
1900	615	100	100	95
2000	615	100	100	100
2100	560	100	100	100

Fig. 14 Chart from Parmiter showing the effect of quenching (hardening) on the stain resistance (comparative efficiency in percent) of chromium cutlery steels. Source: Thum, 1935, p 288

proved, although numerous attempts had been made by adding elements or the slight alteration of elements already present. Parmiter mentions that a modified type had been discovered that had 0.70% carbon and 16.5% chromium, which has higher hardness and is used for cutlery and surgical and dental instruments.

The paper also discussed structure, with micrograph illustrations. There is also a chart showing that carbon steels and stainless steel knives are actually quite close in cutting properties. Information is given on forging, annealing, hardening, and tempering temperatures. A chart (Fig. 14) shows that oil quenching from 1900 to 2100 °F produces steel with high stain resistance. For the compositions he investigated, Parmiter notes: "Within reasonable limits, of which composition plays an important part, the harder the steel, the more resistance the surface is to stain."

Parmiter says that "apart from proper hardening, the most important process by far which influences stainless resistance is grinding." He says, "It should be realized at the start that stainless steel is more difficult to grind than carbon steel. It also dissipates heat very slowly, and great care must be taken in grinding the steel to avoid 'grinder's scorch' and heat checks."

He comments on the remarkable resistance of the steel to nitric acid and the ready attack by both hydrochloric and sulfuric acids.

Stainless Iron and Steel by John Henry Gill Monypenny (1926)

The first treatise on stainless steel that appeared in England, *Stainless Iron and Steel*, was authored by John H.G. Monypenny (1885–

1949) (Fig. 15) and was published by Chapman and Hall, Ltd., London. He was 41 years old when his book was published.

Monypenny studied at University College in Sheffield and had been awarded an Associateship in Metallurgy. In 1926, Monypenny was Chief of the Research Laboratory at Brown Bayley's Steel Works, Sheffield, where Harry Brearley was Works Manager. Monypenny is also acknowledged by Arthur and Harry Brearley in the preface of their book *Ingots and Ingot Moulds* (Longmans, Green and Co., 1918).

Monypenny's 315-page book contains a remarkable amount of information, especially in view of the fact that stainless steels were first produced commercially just 13 years earlier. Four of those years were during The Great War, when little experimental work on stainless steel was possible. The entire output of stainless steel was requisitioned for the manufacture of valves for airplane engines. Monypenny acknowledged the assistance of just four men regarding technical advice, "and in particular, Mr. H. Brearley and Mr. R. Mainprice."

Monypenny stated, "It has been the author's intention to present data pertaining to the various grades of the steel in as clear a manner as possible to aid the user in making a choice of suitable grades for various requirements." The use of the term *stainless iron* in the title was the term originally applied to the low-carbon chromium stainless steels that were not hardenable by heat treatment. Today, it is called ferritic stainless steel. The first commercial heat of that class of stainless had been cast in Sheffield just six years earlier.

Monypenny cites "the modern art of metallography" as being of special importance in the study of stainless steel. He deplored the fact that most people think stainless steel is only for cutlery and some think it is a form of plating. Most of the book is devoted to discussions of the chromium steels, because the chromium-nickel austenitic steels were not made in England until approximately 1920 when the Brearley-Firth Stainless Steel Syndicate and Krupp exchanged patents. There is, however, a very good 20-page discussion of the chromium-nickel steels, including a comparison with the cutlery steel. The Strauss-Maurer diagram (Fig. 16), showing the various percentages of chromium and nickel required to produce an austenitic alloy, was shown.

The phenomenon of carbide precipitation in the austenitic steel was not mentioned. The subject of passivity is mentioned, but Monypenny said he did not fully understand the subject. However, a great deal of discussion was provided on the effect of the chromium content on the resistance to staining and corrosion in various media. The book included over 100 illustrations, many of which are micrographs, and there were 35 tables.

Fig. 15 Prominent contributors in the early developments of stainless steels. Source: Zapffe, 1949, p 22

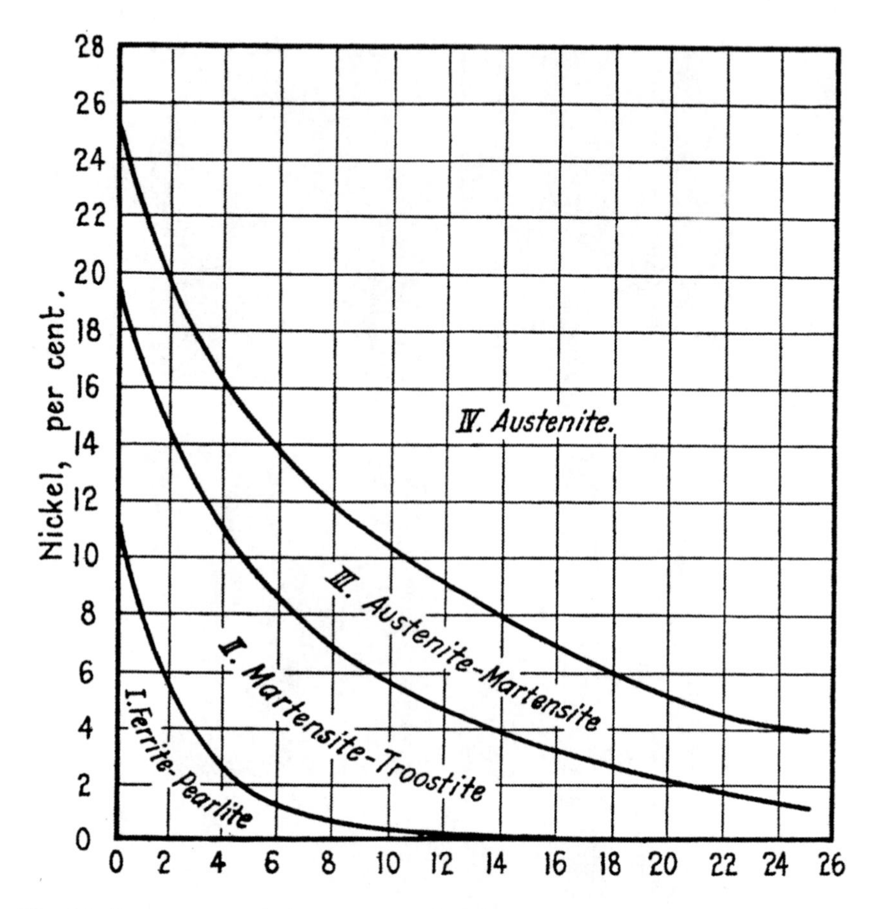

Fig. 16 Percentages of chromium and nickel required to produce an austenitic alloy. Source: B. Strauss, *Non-Rusting Stainless Steels,* Proceedings of the American Society for Testing Materials, 1924, p 208–216

The final chapter discussed applications and cited the use of chromium stainless steels for numerous steam locomotive parts, hydraulic pumps, cutlery, vanes, and blades for steam turbines. Monypenny went on to author six editions of his book, making him the most prolific of all writers on stainless steel.

The Book of Stainless Steels (1933 and 1935)

In 1933, *The Book of Stainless Steels* was published by the American Society for Steel Treating (ASST), which was renamed later that year and became known as the American Society for Metals (ASM).

The book was edited by Ernest E. Thum (Fig. 17). Thum, previously from *Iron Age*, was the first editor of *Metal Progress*, which was a periodical published by the ASST beginning in 1930 and continuing as a publication of ASM until 1986.

Ernest Thum decided to edit a book on stainless steel, written mainly by experts who would each contribute a chapter on subjects that were their specialties. No book on stainless steels had yet been written in America, except for the small trade publication produced by the Firth-Sterling Steel Company of McKeesport, Pennsylvania, in 1923. Thum had a wide acquaintance with experts on stainless steel by virtue of their having contributed articles to *Metal Progress*.

The first edition of *The Book of Stainless Steels* (1933) was very well received, but it was evident that certain subjects should be added and others should be broadened. Thum produced a second edition (1935), which now had chapters contributed by 75 experts covering almost every aspect of the metallurgy of stainless steel. It was an 800-page book that was divided into the following six parts:

- I. General Considerations, which included an introductory and historical chapter by Thum
- II. Production and Fabrication
- III. Properties of the Typical Alloys
- IV. Specialized Tests
- V. Requirements of the Consuming Industries
- VI. Indexes

The detailed table of contents (Fig. 18) is very representative of prominent American metallurgists and industrial companies of the time. It also includes a paper by William Van Alen, who was the famous architect of the Chrysler Building. In his introduction, Van Alen wrote:

> "The Chrysler Building has now been up for four years and is showered with cinders and smoke (at times very dense) thrown off by the New York Edison steam-electric plant, three blocks distant. The metal has not lost its brilliance. It does not collect more soot than a piece of glass, because it has a surface that is as highly polished as glass and has a mirror-like reflective value. This is truly remarkable, as it is not so much a question of durability, but appearance, which in turn is related to various colored deposits which may form on the surface of unchanged metal."

Fig. 17 Ernest E. Thum, first editor of *Metal Progress* and editor of *The Book of Stainless Steels*

Part I — General Considerations

1. HISTORICAL AND INTRODUCTORY NOTE ....
   BY E. E. THUM,
   Editor, *Metal Progress,* Cleveland, Ohio

2. THE GENERAL REQUIREMENTS AND HOW THEY ARE MET ....
   BY JEROME STRAUSS,
   Chief Research Engineer, Vanadium Corporation of America, Bridgeville, Pa.

3. CONSTITUTION OF CHROMIUM AND CHROMIUM-NICKEL STEELS ....
   BY V. N. KRIVOBOK,
   Professor of Metallurgy, Carnegie Institute of Technology

Part II — Production and Fabrication

4. THE PROBLEMS OF MELTING AND CASTING;
   A — STEEL MAKING ....
   BY A. L. FEILD,
   President, Alloy Research Corporation, Baltimore, Md.

   B — FOUNDRY PRACTICE ....
   BY R. D. ALGER AND G. C. McCORMICK,
   Superintendent and Vice-President, respectively, General Alloys Co., Boston

5. ROLLING MILL PRACTICE;
   A — FROM INGOT TO SHEET AND STRIP ....
   BY E. R. JOHNSON AND ROBERT SERGESON,
   Republic Steel Corporation, Canton, Ohio

   B — ROLLING OF STAINLESS PLATE ....
   BY L. M. CURTISS,
   Plate Mill Superintendent, Lukens Steel Co., Coatesville, Pa.

   C — MANUFACTURE OF SEAMLESS TUBES ......
   BY H. D. NEWELL,
   Chief Metallurgist, Babcock & Wilcox Tube Co., Beaver Falls, Pa.

Fig. 18 Table of contents from the second edition of *The Book of Stainless Steels,* American Society for Metals, 1935

Fig. 18 (continued) Table of contents from the second edition of *The Book of Stainless Steels,* American Society for Metals, 1935

Fig. 18 (continued) Table of contents from the second edition of *The Book of Stainless Steels,* American Society for Metals, 1935

Fig. 18 (continued) Table of contents from the second edition of *The Book of Stainless Steels,* American Society for Metals, 1935

Fig. 18 (continued) Table of contents from the second edition of *The Book of Stainless Steels,* American Society for Metals, 1935

Fig. 18 (continued) Table of contents from the second edition of *The Book of Stainless Steels,* American Society for Metals, 1935

Part V — Requirements of the Consuming Industries

Fig. 18 (continued) Table of contents from the second edition of *The Book of Stainless Steels,* American Society for Metals, 1935

Part VI—Indexes

Fig. 18 (continued) Table of contents from the second edition of *The Book of Stainless Steels,* American Society for Metals, 1935

After his review on the range of architectural uses of stainless steel, Van Alen then concluded:

> "In my opinion 18-8 has a wonderful future, for it is probably the most permanent building material known. Atmospheric conditions seem to have no deteriorating effect on it, and I believe that in the not very distant future our steel structures will be covered on the exterior with this material in order to obtain a thoroughly weather and water-proof structure which will require a minimum of upkeep externally. The air-tightness of the walls will save on the amount of heat required for heating. A similar treatment in design of course could be applied to small modern homes."

Overall, the book was an invaluable source of information that became a classic in the industry. In addition to being a collection of technical papers, the book also included an index of over 1000 trade names and other designations of stainless steels and identified each with one of approximately 100 manufacturers. Because the three-digit American Iron and Steel Institute (AISI) numbers for stainless steels had been only recently introduced, none of the authors used these numbers, preferring the old designations, such as cutlery steel, 18-8, and Rezistal 2, and the spelling out of chemical ranges, such as 18 to 20 percent chromium. However, the article "Classification of Trade Names" by Clayton Plummer does mention incorporation of classification adopted at that time by the AISI and the Code Authority for the Alloy Casting Institute (ACI). The AISI and ACI designations mentioned in the index of Thum's book are listed in Tables 1 and 2.

Table 1 Wrought stainless steel AISI designations listed in *The Book of Stainless Steels*, edited by E. Thum (ASM 1935)

AISI Designation	C	Cr	Ni	Other
401	0.08–0.20	9–12		Si
403	0.12 max	12–15		
405	0.08 max	12–15		Al
406	0.12 max	12–15		Al
410	0.12 max	12–15		
411	0.12 max	12–15		Si Cu
412	0.12 max	12–15		Si
414	0.12 max	12–15		Ni
414F	0.12 max	12–15		Ni S
416	0.12 max	12–15		S
418	0.12 max	12–15		W
420	0.13–0.20	12–15		
425	0.12 max	12–15		W
430	0.12 max	16–23		
430F	0.12 max	16–23		S
431	0.12 max	16–23		Ni
431F	0.12 max	16–23		Ni S
432	0.12 max	16–23		Cu Si
434	0.12 max	16–23		Cu Si
435	0.12 max	16–23		Cu Si S
436 (Cr 15–18)	0.13–0.20	16–23		Si
438	0.12 max	16–23		W
439	0.51–0.65	7–11		7–11 wt% W
440	0.13–0.20	16–23		
441	0.13–0.20	16–23		Ni
442	0.21–0.35	16–23		
446	0.21–0.35	24–30		
448 (Cr 18–23)	0.13–0.20	16–23		Si
301	0.13–0.20	12–15	5–7	Mn
302 B	0.13–0.20	16–23	7–11	Si
303	0.13–0.20	16–23	7–11	Mo Se
304 (18–8)	0.07 max(a)	18	8	
305 (19–9)	0.13–0.20	19	9	
306 (19–9)	0.07 max(a)	19	9	
307 (20–10)	0.13–0.20	20	10	
308 (20–10)	0.07 max(a)	20	10	
309	0.13–0.20	24–30	12–15	
312	0.13–0.20	24–30	7–11	
316	0.12 max	16–23	7–11	
320 (18–8)	0.13–0.20	18	8	Ti
321	0.13–0.20	19	9	Ti
327	0.25 max	24–30	3–5	
330		12–15	31–39	
344			16–23	

(a) In the list of alloy trade names of the Thum book, some commercial alloys with higher carbon content (0.12 % max. carbon) were also classified as being AISI types 304, 306, and 308.

Table 2 Cast stainless steel designations listed in *The Book of Stainless Steels,* edited by E. Thum (ASM 1935)

Alloy Casting Industry (ACI) Designations(a)	C	Cr	Ni
C20	Various	31–39	Less than 3%
C–21	Various	24–30	Less than 3
C–22	Various	16–23	Less than 3%
CN–30	Various	24–30	7–11
CN–32	Various	24–30	7–11
CN–34	Various	24–30	16–23
CN–36	Various	24–30	7–11
CN–38	Various	16–23	7–11
CN–40	Various	16–23	7–11
CN–41	Various	16–23	7–11
CN–41	0.13–0.20	16–23	12–15
CN–42	Various	31–39	16–23
CN–43	Various	24–30	24–30
CN–43	Various	31–39	24–30
CN–43	Various	31–39	31–39
NC–1	Various	16–23	67–75
NC–2		12–15	58–66
NC–3	Various	16–23	31–39
NC–4		12–15	31–39
NC–5		7–11	24–30
NC–6	Various	16–23	24–30
NC–7	Various	7–11	16–23

(a) Alloy Casting Industry designation were later superseded by Alloy Casting Institute designations

Stainless Steels by Carl Andrew Zapffe (1949)

In 1949, Carl Zapffe (1912–1994) (Fig. 19) authored the second American book on stainless steel. The 370-page book was published by the American Society for Metals, Cleveland, Ohio. Zapffe, a native of Minnesota, received a bachelor's degree in metallurgy in 1933 from Michigan Tech and a master's degree from Lehigh University in 1934. He was employed at the DuPont Experimental Station, Wilmington, Delaware, from 1934 to 1946. In 1946, Zapffe became an independent research consultant in Baltimore, Maryland. Baltimore was the headquarters of the Rustless Iron and Steel Company and the Eastern Stainless Steel Company.

He was just 37 years old when his book was published. He gave credit to 28 prominent American metallurgists for their assistance in preparing the book. The book was written in the form of a college textbook and included 20 to 40 questions at the end of each chapter. Unlike Thum's book, published 14 years earlier, Zapffe made extensive use of AISI and ACI designations and included the then-current list of those designations (see the inside back cover of this book).

Fig. 19 Carl Zapffe, who authored *Stainless Steels* (1949) and coined the term *fractography* from his original work on the microscopic examination of fracture surfaces. 1968 photo. Courtesy of the Zapffe family

The chapters in Zapffe's book include:

- Historical Background
- Corrosion Resistance
- Metallurgical Constitution
- Martensitic Stainless Steels
- Ferritic Stainless Steels
- Austenitic Stainless Steels
- Production, Fabrication, and Finishing
- Glossary
- Bibliography
- Author Index
- Subject Index

Zapffe, with his 27-page history of stainless steel, has been widely recognized as the preeminent stainless steel historian in this country and abroad. His bibliography has 317 entries. There is an extensive discussion of the phenomenon of passivity as the fundamental basis for the corrosion and oxidation resistance of stainless steels (Fig. 20). However, he also concluded that "the fundamental nature of passivity is still in dispute."

Zapffe was excited about stainless steel and spoke of it as the "miracle metal" and "metallurgy's crowning achievement." The book, even today, is a first-class treatise on stainless steel. He went on to get doctorate degrees in research and engineering. In 1960, he edited a film on stainless steels for the Republic Steel Company. Two of his booklets were published that discussed stainless steel in layman's language and included sketches.

Another major accomplishment of Zapffe was his invention of fractography, which is the term he coined in 1944 following his discovery of a means for overcoming the difficulty of bringing the lens of a microscope sufficiently near the jagged surface of a fracture to disclose its details within individual grains (*Trans. ASM,* Vol 34, 1945, p 71–107). He also discovered that hydrogen embrittlement caused the breaking up of the welded Liberty Ships in heavy seas during World War II. In addition, he became a noted physicist, writing papers critical of Einstein's special theory of relativity.

Zapffe's book was translated into Japanese, an event that led to Japan becoming a major stainless steel producer. This also explains the mystery of why the Japanese adopted the American stainless steel grades and the AISI and ACI designations. Type 304 stainless steel became SUS 304 in Japan's numbering system.

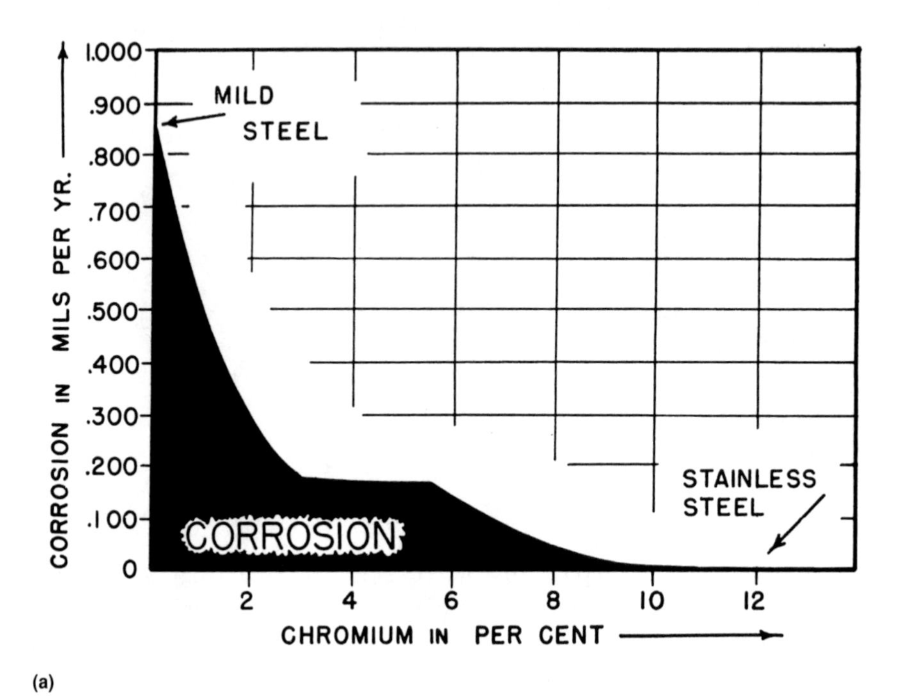

(a)

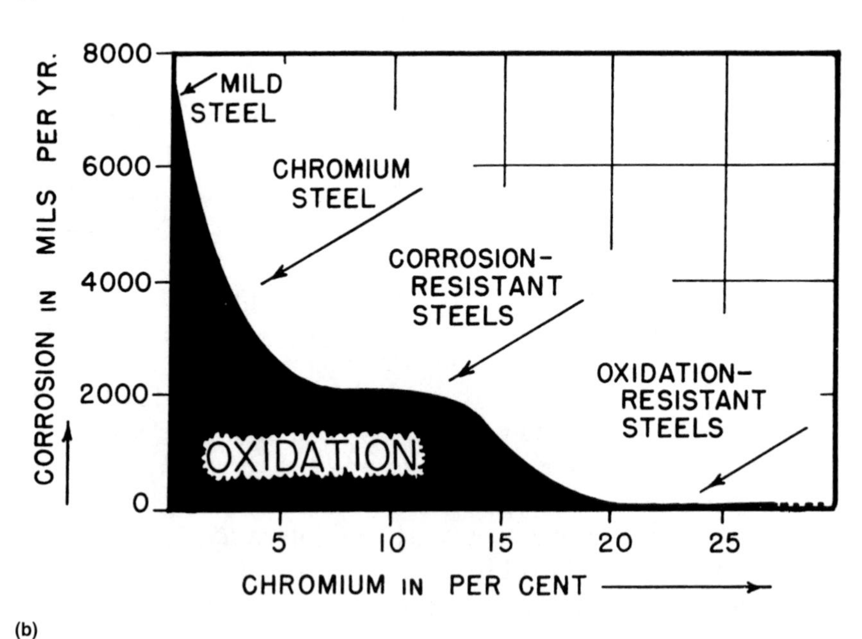

(b)

Fig. 20 Effect of chromium on corrosion and oxidation resistance of steel. (a) Iron-chromium alloys exposed for 10 years to corrosion and rusting in an industrial atmosphere. (b) Oxidation penetration of ½ inch cubes exposed to air for 48 hours at 1000 °C. Source: Zapffe, 1949, p 31, 32

Appendix: Text of 1920 Paper by W.H. Marble

Adapted from "Stainless Steel—Its Treatment, Properties and Applications," *Transactions of the American Society for Steel Treating,* Volume 1, Dec. 1920, p 170–179:

"The demand for a non-rusting or non-staining material for knives and other cutting tools or instruments suggested itself as a big field for Stainless, and this probably was one of its first uses. Scientific tests have since proved conclusively that a Stainless blade properly made, hardened and tempered, will take and maintain an edge equal to the best cutlery, with the added asset that it resists rust and stain.

During the late war period, all Stainless supplies in England were appropriated by the Ministry of Munitions. The bulk of this material was used for a purpose similar to that for which it was designed—exhaust valves for aero engines, which are required to be strong and resist erosion at high temperatures. Its subsequent use in this country naturally followed, and it has been used successfully, and in large quantities, in place of high-nickel and high-tungsten steels.

It has been stated too, that Stainless is "one of the indirect results of improvements in the manufacture of carbon-free ferro-alloys."

Previous Investigations

In 1917, Dr. W. H. Hatfield,[1] of Brown-Firth's Research Laboratory, Sheffield, England, published results of tensile and impact tests of a steel containing about .30% carbon and 13% chromium. He has also shown comparisons between air, oil and water quenching, followed by a relatively high temperature tempering. Some of his results have been incorporated in a report submitted to the joint Iron and Steel Committee of the S.A.E.[2] and A.S.T.M. This report also includes a brief discussion of proper methods of working, annealing and finishing Stainless Steel, with a chart showing the results of scaling tests with Stainless, high-speed and other alloy steels.

Messrs. French and Yamauchi, of the United States Bureau of Standards, show, in Chemical and Metallurgical Engineering,[3] some very

1. W. H. Hatfield, "Heat Treatment of Aircraft Steels," Automobile Engineer, 1917, vol. 7.
2. Journal of the Society of Automotive Engineers, vol. 5, No. 3, September, 1919, pp. 262 and 263.
3. H. J. French and Yoshito Yamauchi, "Heat Treatment of a High-Chromium Steel," Chemical and Metallurgical Engineering, vol. 23, No. 1, July 7, 1920.

valuable tables, giving results obtained by varying the quenching and tempering temperatures. Their results would indicate that quenching from about 1,750° F. develops the best combination of strength and ductility, although this does not give the maximum hardness; that ductility as measured by elongation and reduction of area is very low in those samples quenched from 1,850° F. and above; that maximum hardness is generally obtained at about 1,950° F.; that the most rapid change in tensile properties and hardness occurs in tempering between about 800° F. and 1,000° F.

Messrs. Seidell and Horvitz, at the present time with the New York Testing Laboratory, discuss in *Iron Age,* results of experiments carried out to find the relation between hardness and double carbides in solution. These results show that with a quenching temperature of 2,150° F., a maximum hardness is produced; that after suitable drawing treatment under these conditions, tensile strength and ductility are maximum. They also show by means of photomicrographs that with minor exceptions, the micro constituents of Stainless Steel are analogous to those with which we are familiar in ordinary carbon steel.

Mr. Elwood Haynes, in an address before the Engineers' Society of Western Pennsylvania, November 25, 1919, draws these conclusions as a result of tests made during the past several years:

That an excess of 8% chromium makes the alloys practically immune to nitric acid; that these alloys could be hardened by heating to redness and quenching; that they are distinctly malleable; that the alloy was much more easily worked if the carbon content were kept below 1%. He also states that its immunity to acids is due primarily to its composition; that, although quenching in water enhanced its resistance to a considerable degree, it is best to use oil for quenching, in order to avoid local contractions and stress in the finished product. His tests have shown that air-cooled forgings show a remarkably fine grain and good cutting qualities, depending upon the exact composition and the highest temperature up to which it was heated before hammering, and that, notwithstanding the comparatively high temperatures of working this steel, the bars show almost no scale during the forging operation. By increasing the carbon content of these alloys, they can be rendered sufficiently hard by quenching to scratch glass. Small castings can be made by the use of chill molds.

Chemical Composition

The selection of a standard analysis has been the result of countless tests that showed generally that, within certain limits, stainlessness is increased with the increase of chromium and the decrease in carbon content.

Carbon	.20 to .40%
Chromium	11.40 to 14%
Manganese	not to exceed .50%
Silicon	not to exceed .30%
Sulphur and Phosphorus	as low as possible

The question of increasing the carbon point, as well as the effects of a higher silicon content, have been rather exhaustively investigated, but with no subsequent change, for the present, from the foregoing analysis.

A brief resumé of the effects of certain elements upon the various properties of the alloy may be of interest.

Effects of Various Elements

Carbon.—The carbon content usually found in Stainless Steel is .25 to 0.40%. Under .25% the steel will not harden successfully, while above .40% it is difficult to forge and begins to lose its non-corrosive property. Properly hardened Stainless of correct composition is a martensitic structure with numerous globules of double carbides of iron and chrome present. The steel varies in hardness in direct proportion to its carbon content.

Chromium.—Chrome is the most important element and is practically the only alloy used. It is from the chrome content that Stainless derives its resistance to the various oxides and corrosive agents. Chrome forms carbides with the carbon, which compound not only furnishes the requisite hardening properties, but is in itself one of the most difficult substances known to oxide or corrode. Pure iron is also known to show great resistance to acids and oxidizing conditions, and as hardened Stainless is theoretically a matrix of iron containing chrome carbides, the combination furnishes an alloy which is highly resistant to stain or rust.

Silicon.—Silicon neutralizes the hardening effect of carbon, thereby making possible the use of a much higher carbon content. It has a tendency to remove gaseous impurities and oxides which assists in the formation of a sounder steel mass, free from slag, seams and blow-holes. To offset these desirable properties, however, the silicon retards hardening and increases brittleness. Stainless containing high-silicon cannot be hardened to the same extent as can the same steel without it. It is best then, to keep the silicon content low.

Sulphur and Phosphorus.—Sulphur and phosphorus are injurious in that they assist in the formation of an electrolytic action which induces corrosion. They also have a tendency to segregate. These elements are impurities and should be kept as low as possible.

Manganese.—Manganese counteracts the injurious effects of sulphur by forming sulphides in which formation sulphur does not show a marked tendency towards "red shortness." The manganese content should be kept low, although a small percentage is necessary to assist in hardening.

Other elements such as nickel, copper, cobalt and tungsten have been used in Stainless Steel, but it is doubtful as to whether any great benefits have been derived from their use. The addition of tungsten and nickel so far as it is known, serve only to increase the luster of the polished section.

Manufacturing

The manufacture of Stainless Steel is in a general way very similar to that of high-speed steel. It is melted in two general types of furnaces, known as the Crucible and Electric, the process of operation and construction of which are no doubt familiar to everyone. The steps leading up to the actual heat treatment can be stated briefly.

After the ingots have been cast and allowed to cool somewhat, they are "box annealed" at a temperature as near 1,380° F. as possible, for a period of about 13 hours. They are then allowed to cool slowly to atmospheric temperature. This is most important, for Stainless, like high-speed steel, is air-hardening. Ingots before forgings are usually planed to remove all surface defects. The material will machine very easily after the annealing operation.

Forging and Rolling

Before forging the ingot should be heated slowly up to a temperature of from 1,650° F. to 2,100° F. It should be cogged down by easy stages, care being taken that the temperature does not fall below 1,650° F. If the steel is worked below these temperatures, it is liable to be ruptured or broken under the heavy blows of the hammer. A noticeable feature of this operation is the very inappreciable amount of scale formed. The steel will be hard if allowed to cool in the air after forging, the hardness depending upon the exact composition of the material and the highest temperature up to which it was heated before being finally hammered. This hardness may vary over wide limits—to correspond with a Brinell hardness of 500 if the temperature before hammering was as low as 1,800° F., or to a Brinell hardness of only 250 if the temperature before hammering had reached 2,000° F.

Annealing

One of the characteristics of this steel is that if heated to a temperature of 1,800° F. and allowed to cool quickly, it becomes hardened. Consequently, after forging it must be annealed before it can be machined. Stainless Steel should be annealed about the same as high-grade tool steel, except that the heat should be about 1,400° F. A good way is to heat the steel in pipes, and cool pipe and steel very slowly in a pit, or to heat it up to about 1,400° F. in an open furnace, then close up the furnace and allow the steel to remain there while the temperature drops slowly to 1,100° F. Then take the steel out and allow it to cool naturally in the air. After annealing, it will have a hardness of approximately 200 Brinell and machines easily. Should a softer condition be required, heat to 1,560° F. or 1,600° F., and follow by very slow cooling.

Hardening

This material may be hardened in air, oil or water. A good hardening temperature is somewhere in the vicinity of 1,750° F., quenching in oil or water, or cooling in the air depending on the size or shape of piece treated. The steel should be tempered back to suit the work or results desired as with any other alloy steel. If the higher hardening temperature is chosen, it will require a correspondingly lower tempering temperature—the reverse of a carbon steel.

Temper Colors

Temper colors appear at approximately twice the temperature of ordinary carbon steels, for the higher heats, so that the temper should be drawn to a definite temperature in a bath or furnace instead of drawing to a color. A good tempering temperature for a knife blade is about 280° F. to 300° F., but for a valve it would be necessary to go as high as 1,100° F. These colors are due to very thin films of oxidized metal. As Stainless Steel resists oxidizing, it requires more heat to produce a given color than ordinary tool steel, as is illustrated by the following table, the temperatures of which are only approximate.

Color	Appears on Stainless Steel (Degrees Fahrenheit)	Appears on Tool Steel (Degrees Fahrenheit)
Faint Yellow	. . .	430
Light Straw	575	450
Dark Straw	660	475
Purple (Reddish)	750	525
Purple (Bluish)	850	550
Blue	1,000	575
Gray Blue	1,100	600
Greenish Blue	1,300	625

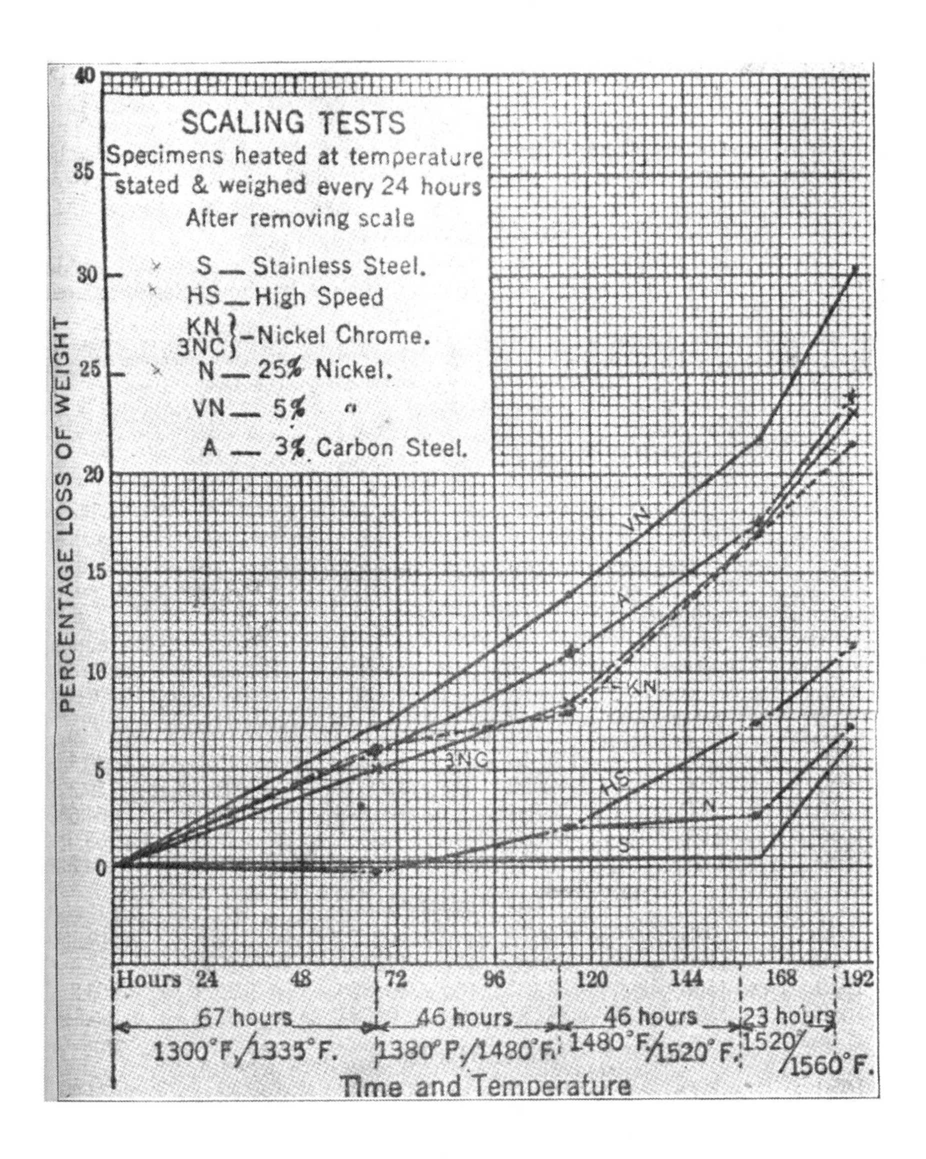

Scaling

As the temperature of ordinary steel rises, the surface oxidizes into a scale of measurable thickness and even at a low red heat, the thickness of the scale increases with time. Stainless Steel behaves quite differently. Up to a temperature of about 1,520° F., the gloss surface due to polishing and hot tinting is permanent, and the specimen neither gains nor loses appreciably in weight. Its comparative value in this respect is illustrated by the following chart which records the percentage lost in weight of various steels after exposure for many hours at temperatures above red heat.

Corrosion

The steel's power to resist stain does not reside in any finish which may be applied by the manufacturer, but in the hardened steel itself. However, rough the surface of a hardened, hardened and tempered article may be, if it is a clean metallic surface and has not been distorted, it will be rustless. However, to exercise this feature to the best advantage, a really good ground polished surface should be provided. Where the highest possible degree of resistance to corrosion is essential, it is recommended that the parts be drawn back very slightly after hardening, and that they then be polished free from scale marks, hair lines or other surface imperfections. These imperfections if left, serve as nuclei from which corrosion is apt to spread. This process is essential only where the bright color of the polished metal is desired to be permanent.

Stainless Steel does not corrode when in contact with other steel. It is, however, attacked in an otherwise non-corroding liquid when in direct metallic contact with copper or copper alloys, such as gunmetal.

Cold working of Stainless Steel destroys its stainless property, and unless it is heat-treated subsequently to restore it, it is useless to put cold-pressed parts into use necessitating its resistance to corrosion. The heat-treatment necessary to restore the stainless properties of cold-worked Stainless Steel is by hardening and tempering. Hardening will necessitate bringing the steel to a temperature of about 1,650° F. Care should also be taken that the material or parts be thoroughly annealed between successive operations.

Physical Properties

The physical properties do not vary greatly when the carbon is within the usual range of composition, or when the steel is hardened and tempered in air, hardened and tempered in oil or water.

A comparison of these properties is shown below, under conditions of air, oil and water-hardened steel of the following composition:

Element	
Carbon	.240
Manganese	.300
Phosphorus	.035
Sulphur	.035
Chromium	12.850
Silicon	.200

Heat Treatment—Mechanical Properties

Hardened from Degrees Fahr.	Tempered at Degrees Fahr.	Elastic Limit	Tensile Strength	Elongation Per Cent	Reduction of Area Per Cent
		Air Hardened			
1650	930	158816	192416	13.0	40.5
1650	1100	99680	120064	21.0	59.2
1650	1300	70784	101248	26.0	64.6
1650	1380	66080	98336	28.0	63.6
1650	1470	70784	96992	27.0	64.7
		Oil Hardened			
1650	930	163072	202720	8.0	18.2
1650	1100	88256	116480	20.0	56.9
1650	1300	77952	105504	25.5	63.8
1650	1380	88256	98784	27.0	66.3
		Water Hardened			
1650	830	158816	202048	12.0	34.2
1650	1100	90272	120736	22.0	59.8
1650	1300	66080	102592	25.8	64.7
1650	1380	67200	97888	27.0	65.2

(The above tests were drawn for one hour at the noted temperatures and allowed to cool in air).

This steel has a co-efficient of expansion of .00001091512, which is slightly less than ordinary steel.

Applications

Cutlery.—Knives made of this steel are in all respects quite equal to those made of the best qualities of cast steel, besides having the advantage of being stainless. The composition of about 13% chromium and .30% carbon is used. Dr. W. H. Hatfield[4] is authority for the statements that Stainless Steel could be hardened far in excess of actual requirements for table knives, carvers, pocket-knives and razors. To get such results, the steel should be hardened from temperatures of 1,740° F. to 1,830° F., or about 250° F. to 300° F. higher than the usual temperatures necessary in the hardening of high-carbon cast steels. Successful tempering can be accomplished by placing the knives in an oil bath for 5 minutes, the oil of which is maintained at a temperature of about 355° F.

Valves.—Valves have been generally made to meet the following specifications, in the annealed condition:

4. Dr. W. H. Hatfield, "Cutlery—Stainless and Otherwise," December 17, 1919, reviewed in the American Cutler, June, 1919.

Yield Point, pounds per square inch	70,000
Tensile Strength, pounds per square inch	90,000
Elongation in two inches	18%
Reduction of Area	50%

The usual heat treatment is to quench in oil from 1,650° F. and temper or draw at 1,100° F. to 1,200° F. One manufacturer hardens valves by heating the previously annealed valves to 1,650° F. and cooling in air. This gives a Scleroscope hardness of about 50. Should a greater degree of hardness be desired it can be obtained by heating to 2,000° F., cooling in air. This gives a Scleroscope hardness of about 75.

Besides its use for cutlery and valves, the purposes to which Stainless can be applied are legion. The fact that nickel steels containing a high percentage of nickel, are difficult to machine, opens up a large field for Stainless, in that it can be readily softened for machining purposes and afterwards suitably hardened to resist wear and stress. As this material is very difficult to deform, and is not easily eroded at temperatures up to 1,500° F., it is an excellent material for the manufacture of hot dies, extrusion dies, punches and moles for die castings.

Other engineering applications of the alloy may be mentioned—turbine blades, pump rods and valves, acid pumps, rams, evaporating pans, races and rollers for ball bearings, refrigerator machinery. Several instances might be cited where turbine and pump parts have maintained their bright appearance, with a very inappreciable amount of corrosion, after having been in use for several years. Due to the extreme toughness, ability to resist abrasian, and its mirror-like finish, moving parts made of this alloy wear longer, and, due to the reduced amount of friction, require less re-packing. Since the steel is affected only slightly by salt water or salt air, it may be employed for a variety of marine purposes.

In the electrical field, Stainless Steel is used for permanent magnets and for electric cooking stoves and utensils, where the advantages of permanently bright and clean surfaces in assisting the reflection of heat and consequent economy of current are immediately apparent.

CONCLUSION

This steel is costly to produce in the form of cutlery by reason of the necessity for better grinding and more careful forging in consequence of its great hardness. For general use, it may be said that it is costly to manufacture by reason of the expensive alloy employed, and the comparatively heavy waste (due to scrap) involved as a result of very careful inspection; but there *is* a very wide field of usefulness before it.

CHAPTER 7

The Chrysler Building (1930)

The Miracle on 42nd Street

"The Art Deco wonder of the world, its eagles still fiercely guarding Lexington Avenue against all incursions of reality, for nothing in the old carnival city appears quite as fantastic as William Van Alen's pulsing automobile-age vision, materializing with all the cathedral-age vision that Walter Chrysler's money could buy."
Claudia Roth Pierpoint, 2004

WALTER PERCY CHRYSLER (1875–1949) (Fig. 21) was born of Dutch ancestry in Wamego, Kansas, on April 2, 1875. His father was a locomotive engineer with the Union Pacific Railroad. After finishing high school at Ellis, Kansas, in 1892, Walter took a job with the Union Pacific as an apprentice in the machine shop, and by the age of 32, he was master mechanic for the entire Chicago Great Western Line.

He moved on to the faltering Buick Motor Company, where he saved the company immense sums by engineering a changeover from wood carriage "bodies by Fisher" to steel body construction. He was eventually the General Manager of General Motors. His earnings sky-rocketed to a million dollars a year by 1920 when Chrysler moved to New York to take over the Willys Overland Company. He was 45 years old. He and his family were soon living on a large French Renaissance estate at Great Neck, Long Island. There was a boathouse and a pier, which led Chrysler to buy a small yacht that he called *Zowie*, which he sometimes sailed to work in good weather.

His thought then was to build his "dream machine," a car that would be far different from any other. He set his engineers to work on

Fig. 21 Walter P. Chrysler

this project, guiding with his own ideas and mechanical skills. By 1923, they were test driving the first Chrysler motor car. He wanted to create a car that would be fast, sleek, and fun to drive. Although the word *streamlined* was not yet invented, Chrysler, in his mind's eye, pictured the car as something that looked fast even when it was standing still.

Chrysler's dream machine was first displayed in January 1924 at an auto show in the lobby of the New York Ambassador Hotel. His cars were soon setting uphill racing records. It was the Roaring Twenties; there were jobs aplenty and a new-found optimism after the postwar doldrums. Building was on the increase, and the sales of motor cars were soaring. An article in *Fortune Magazine* said that Chrysler had made the perfect car for the twenties, "a period when desire supplanted needs."

However, Chrysler harbored yet another idea. He wanted to have his company's headquarters in New York City, and he wanted it to be in the tallest and grandest building in the world. He would build a skyscraper. Chrysler wrote, "I came to the conclusion that what my boys ought to have was something to be responsible for. They had grown up in New York and probably would want to live there. They wanted to work, and so the idea of putting up a building was born. Something that I had seen in Paris recurred to me. I said to the architect, 'Make the building taller than the Eiffel Tower.'"

The First Skyscrapers

The first skyscraper was the ten-story Home Life Insurance Company Building, built in Chicago in 1885. It was made possible with the introduction of mass-produced, long, wrought iron and steel beams and reliable elevators. Until 1888, buildings and other tall structures were built by piling stone on stone or brick on brick. The iron and steel beams were not only long, but they were also strong enough for structural support at greater heights than stone or brick. However, even though the term *skyscraper* obviously refers simply to a very tall building, the architect T.J. Gottesdiener (a partner of the architectural firm Skidmore, Owens, and Merrill) is said to have told the *Christian Science Monitor*, "I don't think it is how many floors you have. I think it is attitude. It is anything that makes you stop, stand, crane your neck back and look up."

In 1913, the 625 foot tall Woolworth Building, the building built with nickels and dimes, was erected in New York City with 57 floors, creating the tallest building in the world, a record that would stand for 17 years.

William H. Reynolds. In approximately 1926, William H. Reynolds, a former New York State senator turned real estate developer, conceived the idea of putting up the tallest office building. New York City was growing and, in the mid-twenties, had surpassed London as the world's largest city. Reynolds bought land on Lexington Avenue between 42nd and 43rd Streets and hired William Van Alen as his architect.

William Van Alen, Architect. William Van Alen was born in Williamsburg, Brooklyn, in 1883, and his career started when he began taking night courses at the Pratt Institute while working as an office boy for some of the best architectural firms. He moved on to some drafting and design jobs. In 1908, when Van Alen was 26, he won the Paris prize for his drawing of "a grand opera house." This led to a three-year stint in Paris at the École des Beaux-Arts and the café Les Deux Magots. In 1911, he returned to New York a much-changed and brash young man who was quoted as saying, "No old stuff for me! Me, I'm new! Avanti!"

Van Alen formed a partnership with a young architect by the name of H. Craig Severance. The men were of completely different personalities, Severance being a very outgoing sort and out on the town most evenings, while Van Alen was more of a loner. After ten years, the partnership finally broke up in 1924 in a dispute over which man had contributed the most to the partnership. Van Alen was 41 years of age.

Each man established his own independent practice. Van Alen completed the designs for several buildings, including a Delman's Shoe Shop where he attracted attention by a novel approach of having the cobblers work inside the windows in full view of passersby. He also received considerable attention for the design of a Child's Restaurant that was built on Fifth Avenue.

Van Alen started working for William Reynolds, who was intent on building the tallest building in the world. He had just finished some preliminary plans for Reynold's building early in 1928 when Reynolds had some financial difficulties and defaulted on his 42nd Street property payment.

Chrysler Hires Van Alen. At this point, Walter Chrysler stepped in and paid $2 million of his own money for the land, the plans, and the architect. In 1928, the W.P. Chrysler Building Corporation was formed, with the following officers:

- Walter P. Chrysler, chairman
- William Van Alen, architect
- Fred T. Ley & Co., Inc., builder
- Ralph Squire & Sons, structural engineers
- H.G. Balcom, consulting structural engineer
- Louis T.M. Ralston, mechanical and electrical engineer
- Harry Arnold, supervising engineer

Van Alen immediately dropped his ideas of adorning the building with glass on the lower floors and a glass dome lighted from within, like a giant glowing diamond. He fell right in with Chrysler's ideas of designing a building that would promote automobiles and the Chrysler Corporation. Chrysler especially wanted to use metals for decorating the exterior of the building. They selected a color scheme of black and white with gray ornamentation and a silvery trim. The style would be Art Deco, which was introduced and named for the 1925 Paris Exposition Internationale des Arts Décoratifs et Industriels Modernes (International Exposition of Modern Industrial and Decorative Arts).

Nirosta (18-8) Stainless Steel

At the Chrysler laboratory, considerable thought was given to this matter of using a metal on the outside of the building and to determine the best alloy for the purpose. A list was drawn up to supplement the information that Van Alen had on the wear and tear on metal store fronts. The potential silvery alloys were listed as follows, in their approximate order of increasing cost:

- Aluminum
- 12% chromium stainless iron and steel
- Benedict metal (57% copper, 28% zinc, 15% nickel)
- Nickel silver (75% copper, 20% zinc, 5% nickel)
- Nirosta metal (18% chromium, 8% nickel, balance iron)
- Monel metal (67.5% nickel, 28.5% copper)

Plated metals were ruled out, because the experience in the automobile industry with chromium- and nickel-plated products in outdoor use showed them to be unreliable. Tests using a milk pitcher made of Nirosta showed that the metal was "stainless as well as rustless, and capable of taking a good polish." Despite the relatively high cost of Nirosta, at approximately 50 cents per pound, Chrysler and Van Alen decided to use this material for the outside of their building. They were sold on its shiny, silvery appearance. The fact that it was a brand new, expensive metal that had never been used on a building made it all the more attractive.

However, the 18-8 austenitic stainless steel was not well known in America. The alloy was first introduced in America during a 1924 American Society for Testing Materials symposium (Chapter 4, "The Great Stainless Steel Symposium (1924)"), when Benno Strauss of the Krupp Steel Works gave a paper describing Krupp's austenitic steel with 20% chromium and 7% nickel. The first production of this alloy in the United States was not until 1927, when several steel companies made some trial heats. However, they were uncertain of what the market would be for an alloy costing over twice as much as the chromium stainless steel, which had been made in America for the past 15 years. The alloy had never before been used in outdoor service or on buildings in this country or abroad. No one could predict whether the alloy would last for five years or fifty in the rather harsh, sooty atmosphere of New York City. Laboratory corrosion tests have never been reliable for predicting the life of an alloy.

Ultimately, three companies, licensees of the Krupp Steel Works in Germany, supplied the metal that was identified by the Krupp trademark Nirosta, a contraction of *Nichtrostende Stähle* (nonrusting steel). The suppliers of sheet were Crucible Steel of America in New York City and Republic Steel in Youngstown, Ohio. Bars and accessories of Nirosta were supplied by the Ludlum Steel Company of Watervliet, New York.

Van Alen's Description of 18-8 Stainless Steel for the Chrysler Building. During the construction and a few years after completion of the Chrysler Building in 1930, papers by Van Alen were published about the architecture and materials of the building. The first was an article in the May 1930 issue of *Metal Arts*. Another was an article in the June 1930 issue of *Metalcraft*. Van Alen also wrote a chapter, "Architectural Uses," in Thum's first and second editions of *The Book of Stainless Steels* (American Society for Steel Treating, 1933, and American Society for Metals, 1935) in which he describes the architectural use of 18Cr-8 Ni (Nirosta) steel in the Chrysler building:

> "As in all skyscrapers, the exterior design is built around extended vertical lines. In the Chrysler Building the color scheme selected was black and white, with gray ornamentation on white surfaces, the trim to be a silvery tone. The use of permanently bright metal was of greatest aid in the carrying of rising lines and the diminishing circular forms in the roof treatment, so as to accentuate the gradual upward swing until it literally dissolves into the sky, the entire composition in design terminating with a final treatment.
>
> "With stainless steel the structural lines and metal facings are intensified by the mirrored surfaces, reflecting the everchanging light from the sky. The splays get black and then brighter as the light reflexes occur, or the position of the observer changes, so that the entire building is constantly changeable, like a brilliant piece of silk waving in the wind. It has all the aspects of brilliant crystal in the sunlight, and a phosphorescent quality when reflecting the moon-light.
>
> "Much experimentation at the motor company's laboratories was done to determine the best metal for the purpose. This was to supplement the mass of information already available to architects about wear and tear on metal store fronts. The silvery alloys available may be listed in approximate order of decreasing cost as monel metal (67.5% nickel, 28.5% copper); 18% chromium, 8% nickel steel; nickel silver (75% copper, 20% nickel, 5% zinc); Bene-

dict metal (57% copper, 28% zinc, and 15% nickel); stainless irons and steels (17% chromium and 12% chromium); and aluminum.

"Cast aluminum has been widely used in large buildings for spandrels because of light weight, price, permanence, distinctive color, and avoidance of drip stain on surrounding surfaces. The straight chromium stainless steels are well adapted for general interior trim where a hard wearing surface is required and where it can be wiped occasionally to retain its original luster, or on the outside where permanence is required and a light tarnish will not be objectionable, like shafts, ties, bolts and other fastenings. Benedict metal is much used in the intricate design of elevator doors, and interior ornamentation, but experience on exteriors was limited, and tests seemed to indicate that it would oxidize in time. Finally 18-8 was tested, first in the form of a milk pitcher. It was stainless as well as rustproof, and was capable of taking fine polish. This 'natural finish' solved the problem, for experience in the automotive industry had established the fact that metal plate of any kind in the present state of the art is somewhat dangerous to use on exteriors.

"An important feature about 18-8 is that it is available in nearly all commercial forms, shapes, and sections, except extruded moldings, and including kalamein, grilles, screens and other fabricated forms. Moldings with sharp corners are made by rolling and drawing. It is fairly easily worked; while it is harder than other white alloys, its superior physical properties enable one to use lighter sections and gages and these are no harder to form, stamp, and otherwise work than the others. It can also be welded readily (an operation frequently necessary in the attachment of cleats for nails or bolts). Its outstanding advantage, however, is that it requires no maintenance.

"Exterior trim for the first few floors of the Chrysler Building was buffed to a high polish. This was done to reflect as much as possible of the light in the somewhat darkened street-canyon. It also is better prepared to resist the more corrosive and dusty atmospheres near the pavement. The same finish was used for store display windows on first floor and basement, door trim, and hardware, as well as in all subway passages. Ornaments in the lobby were worked out as individual pieces and welded together into a whole window frame or door frame.

"Sheet metal on all but the lower part of the building was not ground to as high a luster, but to a satiny gloss. High polish could then be used for relief. Copings, cap and base flashings were of 18 gage metal, ornaments of 22 gage, and roofings of 22 or 24 gage.

"The ornamentation is modernistic in touch, including 9-ft. pineapples (gracing points on the 24th floor), radiator caps with 15-ft. wing spread, and eagle-head gargoyles projecting 9 ft. on the 61st floor. All these were molded into form and the various sheets lock-seamed together. The gargoyles were lined inside with cement mortar, 2 in. thick, reinforced with wire mesh for stability. This was to strengthen them enough to carry concealed flood lighting fixtures.

"The most distinctive part of the design is undoubtedly the roof with its curved sweeps and radiating ribs tapering up to a gleaming 100-ft. spire. All this siding, roofing, and sash is made of 18-8 sheet. Each piece of the metal cresting was cut to individual templet, and joined to its neighbors with lock-seams or standing seams. Some of this metal covered wooden battens; other areas were secured with cleats, nailed, screwed or bolted down with 18-8 fasteners, all completely hidden."

The Groundbreaking and Race

It was 1928, and Walter Chrysler was in the prime of life at age 53. It was the year when Chrysler would bring out his first Plymouth automobile, acquire the Dodge Brothers Company and add the DeSoto to his line, introduce the new Chrysler Silver Dome car, and announce the building of the tallest building in the world. All of this did not go unnoticed by *Time Magazine*, which accorded him the title of "Man of the Year." *Time* cited him as coming from "a motor man with one product, he had been one of the chief industrialists and undeniably the outstanding business man of the year." *Time* quoted him as saying, "I like to do things. I like to build things. I am having a lot of fun!"

The structure of the building would not be revolutionary. There would still be the idea of a steel skeleton structure, but the interior would consist of two sections, including an inner core made of concrete to help stabilize the building in high winds. It would contain the elevators, stairs, and mechanical equipment. The walls would be mostly of brick, with some more decorative stone on the lower three floors, but the walls were not intended to provide support to the building.

At street level, the building would extend for one block on Lexington Avenue, from 43rd Street to 42nd Street, a distance of 205 feet. On 43rd Street, the building would go back the same distance, but on

42nd Street the building went back only 162 feet, stopping just short of a barbershop, the owner of which refused to move. In accordance with building regulations, the building would be set back from the street at five levels, creating, in effect, five separate buildings on top of each other.

The Bethlehem Steel Company in Bethlehem, Pennsylvania, would supply all 20,291 tons of the wide-flange structural steel beams required. A virtually continuous supply of steel beams had to be carefully scheduled, because there was no storage area at the building site. Materials had to be hoisted immediately into the building. The tale is told that sometimes the steel was still warm upon its arrival.

The construction of the Chrysler Building was one of the least documented. Neither Chrysler nor Van Alen left detailed records pertaining to the design and construction of the building. By good fortune, however, David Stravitz, a New York designer, product developer, and amateur photographer, chanced upon negatives that professional photographers had taken during the building's construction. Stravitz discovered the film in 1979 in a closed photographic studio, where the film was about to be sold for its silver content. Stravitz purchased the lot, which consisted of approximately 150 negatives.

In 2002, Stravitz edited a book of approximately 100 photographs of the Chrysler Building during its construction. *The Chrysler Building: Creating a New York Icon a Day at a Time* was published by The Princeton Architectural Press, New York. It has been an invaluable reference and revealed important information about the construction of the tower that was not generally known.

Demolition of an existing building began on October 15, 1928, by the Godwin Construction Company and was completed on November 9. Excavation began a week later and was completed in mid-January when they reached bedrock. Construction began on January 21, 1929. In March 1929, it was announced in the press that the Chrysler Building would be 808 feet tall and surmounted by a sculptured figure 16 feet tall. By April 25, 1929, five floors had gone up, and signs were posted announcing "Fred T. Ley Company, Builder," "Ready for Occupancy Spring 1930," and "Renting Office at 315 Lexington Ave."

The building was right on schedule, and by July 12, 1929, it was announced that it had reached 35 floors. This was followed by an announcement in the press in July that George Ohrstrom, a 34-year-old banker, sometimes called "the kid," was planning to build the world's tallest building at 40 Wall Street. It was said that "the Bank of Manhattan Company Building would reach 840 feet," without mentioning that it would surpass the Chrysler Building by over 30 feet. Then, on August 30, 1929, Al Smith announced the formation of the Empire State Building Construction Company and the plan to erect an office building 80 stories high, or exactly 1000 feet. The building was to be backed by General Motors. Al Smith, now out of office as mayor, would be president of the Empire State Building Company.

The world's first and greatest skyscraper race had begun with three contestants: Bank of Manhattan, Chrysler, and Empire State. Construction of the Bank of Manhattan Building began on May 28, 1929. In the annals of New York real estate, the race for height of 1929 has never been surpassed in intensity and human drama.

On August 30, 1929, the Chrysler Building had reached a height of 55 floors, and Van Alen announced his plans to go higher than the originally planned 808 feet, to a height of 842 feet, a scant 24 inches above the stated height for the 40 Wall Street building. On October 14, 1929, the Chrysler Building had reached 77 stories and, at 861 feet, was 66 feet below the intended height of the Wall Street building. Van Alen, to all appearances, had conceded the race. There was no indication that the scaffolding was being raised to prepare for additional floors.

In the meantime, Ohrstrom and Severance increased the height of their building to 927 feet and celebrated the topping out of the 40 Wall Street building with a lavish party on the first floor in early November 1929. The event was announced in all the New York headlines.

Nine days after the cessation of building at the Chrysler site, on October 23, which was the day before Black Thursday, the day of the New York Stock Exchange market crash, Van Alen and his engineers were standing on the corner of Fifth Avenue and 42nd Street, looking up intently toward the top of their building three blocks away, when they witnessed something like a miracle happening on 42nd Street. A needle, tapered to a point, and a thin, latticed black steel structure was rising ever so slowly up from the scaffolding at the top of the Chrysler Building. Van Alen wrote, "It was like seeing a beautiful butterfly emerging from its cocoon."

The structure continued to rise up out of the building for 90 minutes until it stopped. The building's height had been increased 185 feet to a height of 1046 feet, 4¾ inches, far exceeding the 927 foot height of the Bank of Manhattan Building and even the Eiffel Tower's 1024 feet. Van Alen had performed the greatest trick in the history of architecture. Needless to say, Ohrstrom and Severance, down at 40 Wall Street, were in a state of shock.

Approximately a month earlier, Van Alen had conceived of his fiendish plan and designed the 185 foot tapered, latticed steel tower, which was built in five sections at a shipyard in New Jersey. The sections were then hoisted to the top of the building, lowering the bottom section first so that the others could be loaded on top. The sections were then riveted together to create the tower (Fig. 22), which rose from the interior of the building on October 23.

Scaffolding of steel pipe and heavy netting was erected around the structure so that metal workers could proceed to cover it with stainless steel tiles. The scaffolding was completed on November 19, 1929. Sheet metal shops had been set up on the 65th and 71st floors, where 4500 Nirosta stainless steel tiles 1⁄32 of an inch thick were cut into tiles to exactly cover the entire structure. It was a giant, shiny, metallic quilt (Fig. 23) that weighed 27 tons. To attach the tiles, a thin layer of nailing concrete was first applied to the structure, to which wooden nailing strips were attached. The metal tiles were fastened to the wooden strips with Nirosta screws and nails.

On January 29, 1930, the scaffolding was removed to reveal the shimmering crown and needle—that part of the building which would become the most distinctive and the most admired of the New York City skyline. The exterior of the building was completed on January 29, 1930, just 14½ months after the start of construction. It was said that not a man was lost. Van Alen had conjured up a masterpiece (Fig. 24).

Walter Chrysler Faces a Dilemma. The Wall Street crash came at approximately the same time as the topping-out of the Chrysler Building. It was the worst of times for capitalist bluster, and Chrysler was indeed mindful of the fact that he would have to populate 900,000 square feet with tenants in the largest building in New York to pay off the $7.5 million mortgage. During the 1920s, the average tenancy rate of office buildings had been 50% in the city, but now there was the most serious depression ever.

Chrysler immediately enlisted the services of Brown, Wheelock, Harris, Vought, & Co., one of the city's largest real estate brokers. Brown, Wheelock divided the city into ten separate districts, and ten agents hit the streets armed with brochures extolling the virtues of the building, its size and grandeur, the modern conveniences, and, of course, its fine location a block from Grand Central Station.

Surprisingly enough, by the time the Chrysler Building opened in May 1930, tenants had been found for 70% of the building, with a list of prominent corporations such as General Electric, Western Union, Time Magazine, Fortune Magazine, and, of course, Chrysler Motor Car.

Fig. 22 Latticed framework of the spire and needle of the Chrysler Building. After the lattice was raised from within the building, a scaffold was built around it, so that workers could start affixing the stainless steel panels. Reprinted with permission from David Stravitz, Ed., *The Chrysler Building: Building a New York Icon a Day at a Time,* Princeton Architectural Press, NewYork, 2002

Fig. 23 Stainless steel sections on the Chrysler Building. Photo taken on June 20, 1930, by the builder Fred T. Ley & Co., Inc. Reprinted with permission from David Stravitz, Ed., *The Chrysler Building: Building a New York Icon a Day at a Time*, Princeton Architectural Press, New York, 2002

On the ground floor, the Chrysler Automobile Salon would occupy the corner of the building at Lexington Avenue and 43rd Street, and at the corner of 42nd Street there would be a Schrafft's Restaurant. A travel agency would be on the left of the imposing three-story main entrance on Lexington Avenue. On the right of the entrance there was to be Florsheim's Shoes and Adam's Hats. However, the prized tenant would be the Texas Company, now Texaco, which rented 14 entire floors. One provision of the Texas Company's tenancy was that there must be a restaurant where their executives could eat lunch—an idea that lead to the creation of the Cloud Club.

Despite its being the 1930s, the Chrysler Building maintained a high occupancy rate, and the $7.5 million in mortgage indebtedness was paid off in 1937, 11½ years early.

Fig. 24 Chrysler Building (same photo as cover). Courtesy of Catherine M. Houska, TMR Stainless, and the Nickel Development Institute

Chrysler's rivals were not so fortunate. The Empire State Building had very few tenants throughout the 1930s and was derided with the nickname "Empty" State Building. The Bank of Manhattan Building at 40 Wall Street went bankrupt in 1935 and was sold for only $1.2 million, approximately the cost of its elevators.

Opening Ceremonies

The opening ceremonies for the Chrysler Building at 405 Lexington Avenue were held on May 28, 1930. Over the main entrance on Lexington Avenue was a huge facsimile of the wings of a Chrysler radiator, and set between the wings was the base of a flagstaff that extended out over the street. The first things to catch the eye in the lobby were the bright red walls of Moroccan marble, the yellow Siena marble floors, and a ceiling completely covered with a canvas by Edward Trumbull entitled "Energy, Result, Workmanship, and Transport," a frieze depicting buildings, airplanes, and scenes from Chrysler assembly lines. The graceful marble stairs to the mezzanine and the basement had glimmering stainless steel rails.

Circulars were passed out that gave the building's statistics and advantages, "providing every contribution to efficiency, sanitation, comfort that human ingenuity can conceive or that money can buy." The building had an area of 900,000 square feet. There were 32 elevators, including the private elevator that served the Cloud Club. When fully occupied, it would house 10,000 tenants and employees, not to mention thousands of visitors. The custodial staff would number 350, and the building would be cleaned daily with a modern central vacuum cleaning system. The building cost $14 million. It was further noted that the building was served by street cars on Lexington Avenue and 42nd Street, a subway line just beneath the building, and a below-ground concourse to the Grand Central Station just a block away.

The Observatory. For 50 cents, visitors could take an elevator to the observatory on the 71st floor. The elevator had beautifully carved wooden doors and interiors styled like Parisian drawing rooms. Ascending took a seemingly endless minute. Inside the observatory was a glass case that displayed Walter Chrysler's fine set of tools, which he had made and used when he was an apprentice machinist. It was Chrysler's way of dramatizing his rise from a lowly machinist to the owner of the world's tallest building.

The observatory was two stories tall, and the views in every direction were breathtaking. On a clear day, it was possible to see 100 miles.

To the west, it was the Hudson River, New Jersey, and the intervening three tall stacks of the Edison Electric coal-fired steam plant just three blocks away. In another view, the Statue of Liberty could clearly be seen, as well as the loser of the skyscraper race, the Bank of Manhattan Building at 40 Wall Street. The steel skeleton of the Empire State Building was just two blocks away, where workmen could be seen perched precariously on beams 500 feet above the street. The Empire State Building would surpass the height of the Chrysler Building when it opened the following year.

The Cloud Club, which occupied floors 66, 67, and 68, just below the observatory, was open only to members and their guests. The membership was capped at 300, and women were not invited. Because of the City Planning Board's requirements, the building had five "setbacks" that gradually reduced the size of the building as it approached the top. This accounted for the fact that the Cloud Club required three floors.

The lower floor of the club, in a baronial style with oak paneling, had a Grill Room, a cloak room, and what was reputed to be the grandest of all men's rooms in the city. On the same floor, there was a kitchen, a stock ticker room, a barber shop, and a room with private lockers, identified only by pictograms, where members could stash their bottles during Prohibition.

From the lower floor, a grand stairway led to the Cloud Room, a room with a high, dome-shaped blue ceiling painted with clouds. The room was very modern, with four large mirror-covered columns and lights at the top, covered with etched glass. The chairs and tables were made of aluminum, which was popular for Art Deco furniture in those days. The seating capacity was 80. One final touch was the possibility of receiving calls by plugging a telephone jack into one of the wall outlets.

A small staircase led to the third floor of the Cloud Club and the private dining rooms of the Texas Company and the Chrysler Company.

The Cloud Club eventually accepted women guests. It closed in the late 1970s because of waning membership. New and larger "cloud clubs" had been built around the city.

Van Alen's Vision

Van Alen's distinctive crown "with its curved sweeps and radiating ribs tapering up to a gleaming ten-foot spire" (Thum, 1933) is undoubtedly the iconic image of the building itself. It is also an icon of skyscraper architecture and the New York City skyline. In 1996, Judith Dupré, a renowned curator, writer, and lecturer, published a tall, thin book entitled *Skyscrapers: A History of the World's Most Famous and Important Skyscrapers* (Black Dog & Leventhal, New York, New York, 1996). She singled out the Chrysler Building for the cover of her book, which clearly showed the stainless-steel-clad tower.

> "The prima donna of all skyscrapers—the extravagantly-topped Chrysler Building—remains the belle of the New York Skyline. Its flamboyance reflects the fevered pitch of the roaring twenties, a time when New York was ignited by the rhythms of jazz and in love with itself as the zenith of urban chic."
> Judith Dupré, 1996

The imagery and boldness of the Chrysler Building definitely evokes wonder:

> "Other towers may claim to being taller now, or being more innovative, but the Chrysler top is a magical sorcerer's wand, a phantasmagoric pinnacle worthy of the Land of Oz. There were the pyramids, the Colossus of Rhodes, the great Medieval and Gothic cathedrals of Europe, and in 'modern' times only the Eiffel Tower, the Woolworth Building and the Chrysler Building: man-made monuments that transcended the parochial vision to inspire delirium and phantasy."
> Carter B. Horsley, *The Midtown Book*

However, at the time of the opening, the Chrysler Building was not greeted with enthusiasm, at least by those who wanted to express an opinion. Lewis Mumford, a famous critic, called it "commercial advertising." William Van Alen was given the unflattering title of "the Ziegfeld of architecture." Nonetheless, the day would come when the building, if not the tallest, would be recognized as one of the world's finest.

When the building was finished, Chrysler at first did not pay Van Alen his fee, claiming that he had taken kickbacks from some of the suppliers. Van Alen filed a $725,000 lien on the building for the balance of his fee. Although no contract had been signed (because Chrysler had taken over the leasehold on the project from the previous owner), the courts prevailed in Van Alen's favor, handing down its highest judgment. The $840,000 fee for preparing plans and specifications, and for closely supervising construction, was calculated at 6% of the $14 million cost of the building. This was considered a bargain, in view of the fact that other architects usually demanded a 10% fee.

It was the depth of the Great Depression, and Van Alen apparently never had another substantial commission. One did not sue *Time's* "Man of the Year" and hope to get many commissions. Van Alen attended the architect's Beaux Art Ball with his wife in 1931. All of New York's architects were there, each outfitted with a headdress of one of their own buildings. Van Alen wore a beautifully made, three-foot-tall headdress of the Chrysler Building's tower, with only five of its seven arched domes, and was dressed as a harlequin.

Walter Chrysler died in 1949 at the age of 74, and the family sold the building in 1953. Because little was done about repairs, the building had fallen into disrepair, and there were few tenants by 1975, when it was acquired by the Massachusetts Mutual Insurance Company of Springfield, Mass. Massachusetts Mutual paid approximately $35 million for the building and spent another $23 million on restoration. The murals in the lobby were restored, and the stainless steel eagles and the spire were polished. Even more important was outfitting the building with a modern heating, ventilating, and air conditioning system. Having achieved this, the building was sold again in 1979. By then, the building was 96% occupied.

The building received landmark status in 1978. In 1981, the interior of the tower was lighted with fluorescent tubes for the first time, so that the 30 triangular windows on each side were like bright stars, making the Chrysler Building one of the most recognizable on the New York City skyline after dark.

The Exterior

Besides the iconic crown, the Nirosta stainless steel was an integral part of Van Alen's vision, as he notes in his chapter "Architectural Uses" in *The Book of Stainless Steels*, edited by Thum (1933, 1935):

> "The use of permanently bright metal was of greatest aid in the carrying of rising lines and the diminishing circular forms in the roof treatment, so as to accentuate the gradual upward swing until it literally dissolves into the sky."

Many ornaments of the Chrysler Building also were created of Nirosta stainless steel. The outside of the building had many heroic ornaments, approximately 50 in all, which jutted out from the four corners of the building at five levels like gargoyles from Gothic cathedrals.

There were nine-foot-tall pineapples, the symbols of hospitality, on the 24th floor. Stretching out 9 feet on the 61st floor were fierce-looking bald eagles (Fig. 25) with 15 foot wing spreads. Margaret Bourke-White, then an aspiring young photographer, had a *Fortune Magazine* office on the 61st floor. She crept from a window with a camera to sit daringly behind one of the eagles while her assistant took the famous picture of her that would appear in *Life Magazine* one day.

Closer to the street, brick girdled the building. Outsized tires pictured in black brick were centered with large-scale, shiny 1929 Plymouth hub caps. All of these ornaments were created of Nirosta stainless steel.

Stainless steel producers also had obvious interest in the Chrysler Building as a bold experiment on the durability of stainless steel in architectural use. Committee A-10 on Stainless Steel of the American Society for Testing Materials had been organized in 1929. Members of that committee viewed the Chrysler Building as a fine opportunity to study the effect of the environment on this new type of stainless steel. A small committee was set up, comprised mainly of stainless steel producers, to inspect the stainless steel panels on the Chrysler tower every five years and determine their condition. This was done for 30 years, until 1960, at which time the inspections were discontinued because there had been virtually no deterioration of the stainless steel panels.

The image of the Chrysler Building continues. In approximately 2000, three major stainless steel producers published some handsome company brochures that enumerated the advantages of stainless steel. Interestingly enough, every company had featured the Chrysler Building on their covers, demonstrating the fact that this building is the very icon of the stainless steel industry. The three companies were licensees of Krupp Thyssen, the discoverer of the alloy in 1912 at the Krupp Steel Works in Essen, Germany:

- Ludlum Steel Company (bars and accessories), Watervliet, New York
- Crucible Steel Company of America (sheet), New York City
- Republic Steel Company (sheet), Youngstown, Ohio

The ornaments, copings, and flashings were fabricated by Benjamin Riesner, Inc. of New York City.

Tower Inspection in 1995. In 1995, following a cleaning of the building, the tower and gargoyles were inspected by Ms. Catherine M.

Fig. 25 Nirosta (18-8) stainless steel eagle on the 61st floor of the Chrysler Building in winter 1929–1930. Photo taken by the famous photographer Margaret Bourke-White, whose studio was just behind the eagle. Reprinted with permission from David Stravitz, Ed., *The Chrysler Building: Building a New York Icon a Day at a Time,* Princeton Architectural Press, New York, 2002

Houska, a senior engineering consultant at TMR Stainless, Pittsburgh, on behalf of the Nickel Development Institute in Toronto. The last inspection had been in 1950. Ms. Houska's report is as follows:

"Mankind has always built large structures out of durable materials as a means of expressing power and wealth. Skyscrapers are the pyramids of yesterday. So, it is fitting that the first large applications for stainless steel were in the tallest buildings in the world: The Chrysler (1930) and Empire State (1931). Although the Chrysler Building was only the tallest building in the world for a few months (77 floors, 1,049 ft or 319 m), its elegant, art deco styling has made it an enduring, internationally recognized example of excellent skyscraper design.

"The Chrysler Building's height ensures that the top portions are cleaned by wind-blown rain at near hurricane force levels during heavier rainstorms. This frequent cleaning has meant that the Nirosta stainless steel exterior, a precursor to Type 304, has performed much better than would normally be expected in a coastal location that was exposed to high pollution levels during much of the structure's life. The top of the structure has only recently been manually cleaned twice, in 1961 and 1995.

"During the final stages of the 1995 cleaning I inspected the Chrysler Building for the Nickel Institute, and Roy Matway separately inspected it for J&L Specialty Steel. We both found the steel to be in excellent condition. The surface deposits had been successfully removed and the relatively high reflectivity 2B finish of the roof looked like it had just been installed. A few dents were evident, probably from scaffold impact during cleaning. Some joints on the spire and gargoyles had cracked and were re-soldered when cleaned.

"A few decorative, 0.019 inch thick, battens, near the top of the spire had to be replaced, because of intergranular corrosion. The high surface phosphorus concentrations on these battens indicated that the problem was probably caused by inadequate flux removal after the initial soldering. Matway obtained two small roof samples and determined that they were within the standard chemistry range of Type 302.

"Some superficial corrosion pitting on balcony flashing (above the top gargoyles) and surface pits up to 0.005 inch deep were seen above the building's flue (east side). The pitting above the flue was discovered during the first cleaning in 1961 when the

> building's incinerator was still in use. At that time, the pitted area was coated with roofing tar and painted silver. In 1995, the silver paint was long gone but much of the roofing tar was still on the surface. The building owner was advised to have the tar removed and the surface abraded and buffed to remove the shallow pits. This was not done until a few years later when a significant effort was made to seal the roof seams, which had always leaked during heavier rainstorms."

In 1997, Tishman Speyer Properties took over and did further restoration.

The Chrysler Building at 75. On Thursday, May 26, 2005, *The New York Times* published a special 14-page section of the paper devoted to the story of the Chrysler Building. This was perhaps the only building to have been so recognized on its anniversary.

The first page of Section F had an eye-catching, full-page photograph of just the top 20 feet of the crown and spire of the building, so that the individual stainless steel panels could be clearly seen surrounding the triangular windows. There were 22 articles, which were illustrated by dozens of photographs. Some of the articles discussed certain parts of the building, while others were the fond remembrances of the writers' visits to the building.

Michael J. Lewis, a professor of art history at Williams College, wrote a long article entitled "Dancing to New Rules, A Rhapsody in Chrome." He wrote the following after describing the lower part of the building:

> "The culmination of this is the building's crown, which is composed of seven radiating arches outlined with zigzag patterns, which create an effect like a cascade of fireworks, mounting upwards in quick succession. Its sense is one of discordant elegance, and perhaps comes closer to what George Gershwin achieved in music than any of Van Alen's contemporaries, a rhapsody in chrome."

REFERENCES

- Thum, E.E., Ed., *The Book of Stainless Steels,* 1st ed., The American Society for Steel Treating, 1933
- Thum, E.E., Ed., *The Book of Stainless Steels,* 2nd ed., American Society for Metals, Cleveland, Ohio, 1935

CHAPTER 8

Edward G. Budd (1870–1946), Inventor and Entrepreneur

"It would seem, therefore, that the whole railway industry is a fruitful field for widespread application of corrosion resisting materials, and that the present time is most propitious for efforts in that direction."
E.J.W. Ragsdale, 1935

EDWARD GOWAN BUDD (Fig. 26) was born on December 28, 1870, in Smyrna, Delaware. From an early age, he was intrigued with all things mechanical. He was destined to become one of the outstanding men in the history of stainless steel, as a developer and builder of stainless steel trains, aircraft, truck trailers, and other products.

The Early Years

When Budd graduated from high school, he apprenticed as a machinist at the G.W. & S. Iron Works in Smyrna. In 1890, when he was 20, he moved to Philadelphia and took a job at the Sellers Machine and Foundry Company, where he stayed for nine years. During that time, he took night classes in drafting and mechanical engineering at the Franklin Institute, the University of Pennsylvania, and the International Correspondence School.

In 1899, when Budd was 29, he took a job at the Niles-Bement-Pond Company, a machine tool company. He was married the same

Fig. 26 Edward Gowan Budd. Courtesy of the Hagley Museum and Library

year and moved to a job at the American Pulley Company as their chief draftsman. The company stamped steel pedestals for Hale & Kilburn, a Philadelphia firm that produced seating for railways and street cars.

Budd Builds the First Steel Railway Carriage Seating. Budd was hired away from American Pulley by Hale & Kilburn. Here, he first appears to have an opportunity to use what will become a life-long creative ability. He was soon developing pressed steel replacements for their wooden railway carriage seating, in recognition for which he was appointed Works Manager. Throughout his career, Edward Budd will become noted for building "the first" of many products. Substituting steel for wooden railway carriage seating was just the beginning.

Budd Builds the First All-Steel Railway Car. Hale & Kilburn then got into the railway carriage construction business. Railway "carriages" were made of wood, which was somewhat flimsy and quite dangerous. In those days, the carriages were heated with cast iron stoves, and tragic fires were not uncommon. However, Budd dreamed up an entirely new scheme of carriage-making and designed the world's first all-steel railcar. Budd's managers at Hale & Kilburn approved the construction of one steel railcar as a trial. The car was lighter, stronger, and significantly more fire resistant than any carriage ever built. When people from the Pullman Company saw the steel car, they were so highly impressed that orders soon began pouring in.

About that time, the Hupp Auto Company of Detroit placed an order with Hale & Kilburn for steel stampings for automobile bodies. This was all that Budd needed to realize there could be a tremendous market for stamped steel auto bodies in the burgeoning automobile business.

The Automobile Body Business

Budd left Hale & Kilburn, and with $75,000 of savings and $25,000 borrowed from two friends, he set up the Edward G. Budd Manufacturing Company and leased an office in the North American Building at 121 South Broad Street in Philadelphia. It was 1912, and Budd was 42 years old. He brought two friends with him from Hale & Kilburn: Joseph Ludwinka and Russell Leidy. Ludwinka was an Austrian immigrant who was extremely talented in the working of sheet metal and had the rare ability to design and make the intricate dies and punches needed for metal stamping and pressing. Budd was confident that, between his own abilities and those of Ludwinka, they could have a successful business making steel automobile bodies. How right he was!

The first Budd plant was at Tioga and Aramingo Avenues in the Kensington section of Philadelphia. It was 1913. During the first year, there were orders for auto bodies from Packard Motor Car, Oakland Motor Car, and for some Peerless truck bodies, but nothing substantial. A year later, Budd moved to a larger facility at Ontario and I Streets. They purchased their first sheet metal press and stored their supply of steel sheet in a circus tent in an adjoining lot. Their first big order came when the newly formed Dodge Brothers Motor Company ordered 5000 bodies for their new "Touring" model. The touring car was so successful that a second order was placed—this time for 50,000 auto bodies. Dodge would remain Budd's biggest customer until 1925, when Chrysler bought the Dodge Company.

In 1914, the Budd Company received a patent for the world's first all-steel, all-welded automobile body.

By 1915, when the company was just three years old, Budd had 600 employees, and they moved again, this time to their permanent location at Hunting Park and Wissahickon Avenues in the Nicetown section of Philadelphia. Production increased from 100 car bodies a day to 500.

Readers may wonder how it is possible for a company in Philadelphia to make a business of selling auto bodies to companies 700 miles away in Detroit. The answer is that the Budd Company had the powerful presses, the know-how to produce the dies, and the ability to select and form the steel. The making of car bodies involves not just stamping out parts but also the fine art of what is called the deep drawing of metal, which is a stretching of the metal until it almost breaks. Budd had a virtual monopoly in this field. In time, the automakers did enter the business, but the Budd Company produced approximately 40% of the auto bodies in America for many years.

One year later, in 1916, Budd produced the 100,000th touring car body for Dodge.

In 1916, Budd also founded the Budd Wheel Company in Detroit, which made wire-spoked wheels with steel rims instead of wooden wheels, another "world's first." In 1924, the Budd-Michelin Company, a joint venture with the French Michelin tire company, was set up in Detroit to make improved wheels for trucks. Budd made the steel disc wheels, and Michelin supplied the heavy-duty tires.

In the 1920s, sales grew unbelievably well, and orders were received from American and foreign companies, including General Motors, Chrysler, DeLage, Citröen, Morris, and Nash.

Budd now owned 75 acres of land and a dozen buildings that occupied an entire city block on Hunting Park Avenue between Fox Street and Wissahickon Avenue. He was sometimes referred to as "the Henry Ford of the East."

The production of thousands of car bodies did not occur without difficulties. There were sometimes problems with the steel that came from the Sparrows Point Plant of the Bethlehem Steel Company in Baltimore. There were the visible problems, such as scratches, rolled-in scale, and rust. However, there were more complex problems, such as when some lots of steel, when deep drawn into parts, exhibited an unacceptable rough surface that was known in the trade as "orange peel." Some of the formed pieces had wormlike surface defects that were called stretcher strains, which would not be completely hidden by subsequent painting. The steel mill metallurgists did not always know exactly why these defects showed up, but they did their best to deliver a good product and cooperated fully with the metallurgists at Budd. The art of steelmaking was still on a learning curve. With steel orders of approximately 20,000 tons a month, Budd was Bethlehem's biggest and best customer.

To cope with these metallurgical and other problems, Budd began to build a fine staff of engineers and a metallurgical laboratory. Two metallurgists were hired in the mid-1920s: Joseph Winlock and Dr. Ralph W.E. Leiter, who would discover how the speed of a press, die design and lubricants, and variations in steel quality could make the difference between a broken part and a perfect one. They worked constantly with the mill metallurgists in an effort to improve the steel. Colonel Earl James Wilson Ragsdale, formerly an Army officer and a brilliant mechanical engineer, was hired. He would become an expert at electric resistance spot welding, which was the way the pieces of an auto body were joined. He developed automatic equipment that almost eliminated the possibility of bad welds. This was most fortunate, because it is impossible to know if a weld is good by simply looking at it.

A New Kind of Stainless Steel Arrives in America

Although the 18-8 type (18% chromium and 8% nickel) of stainless steel was discovered at the Krupp Steel Works in Germany in 1912, this new alloy that did not rust was not made in America until 1928, when the Allegheny Steel Company and the United States Steel Company began its production, having obtained licenses from Krupp. It was not known at first who their customers may be. It was a material with many unique properties, but it was more difficult to produce than carbon steel, and it cost 15 times as much.

Budd metallurgist Joseph Winlock thought the company should look into this 18-8 stainless steel, especially because, unlike other steels, the material could be cold rolled to extremely high strengths and still retain enough ductility to permit forming it into structural shapes. The idea was to build a very strong, lightweight structure with the use of thin material. However, there wasn't a welding method that also would not anneal the high-strength steel in the area adjacent to a weld. There was a second problem. Welding created a condition called carbide precipitation in the area adjacent to a weld, a phenomenon that would lead to the loss of corrosion resistance in the area adjacent to the weld.

However, the Budd engineers, who were experts in the art of electric resistance spot welding, which is done much more rapidly than other types of welding, thought that the speed of spot welding may eliminate the problems of annealing and loss of corrosion resistance.

After obtaining several sheets of the cold-rolled 18-8 material, some spot welding tests were made using the regular settings for the welding of car bodies, which included an approximately one second weld time. It was found that a determination of carbide precipitation could be made using the Strauss test, which required the immersion of the welded sample for 72 hours in a boiling solution of copper sulfate and sulfuric acid. The test samples failed the test, because the acid had dis-

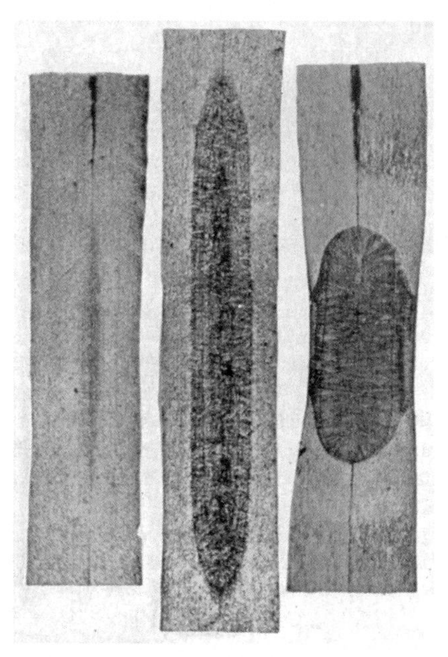

Fig. 27 Cross section of shot weld specimens of 18-8 stainless steel with (left) inadequate fusion due to insufficient heating, (middle) correct heating and fusion, and (right) excessive heating with carbide precipitation in the heat-affected zone of the weld. Source: E.J. Ragsdale, To Weld 18-8 Minimize Time at Heat, *Metal Progress,* Feb. 1933, p 26

solved the steel in a circular groove around the weld. Over the next few months, hundreds of weld tests were made using different settings until, finally, the right combination of settings of amperage, pressure, and welding time was discovered to ensure proper fusion (middle specimen in Fig. 27) while also not overheating to cause carbide precipitation and a reduction of corrosion resistance. The welding time needed to be on the order of 1/100 of a second.

Earl Ragsdale's Shot Weld Patent

On March 14, 1931, Ragsdale filed for a U.S. patent on "Method and Product of Electric Welding." The application showed that Budd was interested only in the electric resistance spot welding of stainless steel sheets with strengths of 150,000 to 250,000 pounds per square inch—strengths that were three to four times the strength of structural steel. It was a ten-page, skillfully written document with a page of diagrams. It carefully explained the reactions within the steel that occurred during welding, a subject that few metallurgists then or now fully understand.

The following excerpts are taken from the patent application:

United States Patent Office
1,944,106
Method and Product of Electric Welding
Earl J. W. Ragsdale, Norristown, Pa,
Assignor to Edward G. Budd Manufacturing Co.,
Philadelphia, Pa
Filing date March 14, 1931 Patent granted Jan. 16, 1934

"By reason of their attainable strength and resistance to corrosion, these steels lend themselves admirably to structural uses and especially to structures which are exposed to the weather or subjected to considerable vibrations, and more especially where great strength with minimum weight is an important factor, such for example, as in aircraft structures. These steels also have the valuable property for some structural uses of being substantially nonmagnetic.

"Because of their valuable characteristics for structural purposes, and because of the unsatisfactory character of many other metals for many such purposes, there has been for some time a great demand for the use of these steels for many structural purposes, but prior to my invention it was not known how these steels could be welded without destroying or substantially impairing, at the welded portions of the structures, the inherent valuable characteristics of the parent metal for structural purposes.

"Prior to my invention this steel has found practically no commercial field of usefulness in the cold worked state which is the state in which it possesses the most valuable characteristics for structural purposes, and this lack of use was largely if not entirely because of the difficulty of satisfactorily joining the several parts made of any structure made of this steel.

"Riveting of this steel is very difficult and hazardous where strength is required. The welding of this cold worked steel is also hazardous for the following reasons: First, because the steel is annealed by air cooling when its temperature is brought to a temperature below its melting point, and if too much of the area of the steel is allowed to become annealed it weakens the structure at the joined areas due to the fact that the strength of the annealed steel is considerably below that of the cold worked steel, and secondly, because there is a temperature range of the metal at which carbide precipitation may occur, and this carbide precipitation will result in destroying the resistance of the metal to corrosion as well as its homogeneity. This latter feature is true as to both the cold worked and the annealed stock. This invention overcomes these difficulties, as will be later described."

It is particularly interesting to note that Ragsdale said, " . . . prior to my invention it was not known how these (high strength) steels could be welded without substantially impairing, at the welded portion of the structure, the inherent valuable characteristics of the metal for structural purposes."

He goes on to explain that his method prevents the occurrence of carbide precipitation, which destroys the resistance of the metal to corrosion. Ragsdale's method preserves the high strength and high fatigue strength by the prevention of annealing, and he avoids the formation of carbides that would reduce the corrosion resistance. This is all possible, as explained elsewhere in the patent, because the welding time is automatically controlled to have a duration of only 1/100 of a second.

Ragsdale does not explain why the formation of carbides would reduce the corrosion resistance, but one commonly accepted theory is that when the temperature of the metal is in the range of 850 to 1500 °F, at which temperature the metal is solid, chromium and carbon atoms combine at the crystal boundaries to form chromium carbide particles. The metal in solid solution near the crystal boundaries is thereby reduced in the chromium content, causing a loss of corrosion resistance.

The Budd Company received U.S. Patent 1,944,106 for the shot weld process. This would prove to be one of the most important patents in the history of the Budd Company.

The World's First Stainless Steel Airplane—The Pioneer

Some time in 1931, the Budd Company began building a stainless steel airplane. To many, this sounded like a foolhardy project, for certainly it was not believed that there was a market for such a machine. However, Mr. Budd was desperate. He wanted to do something that would attract attention and show the world what his company could do with the new metal, high-strength stainless steel, which would be fabricated using the Budd shot weld process.

The Great Depression had begun late in 1929, and orders for automobile bodies had declined by ⅔. A substantial part of the 10,000-man work force had already been laid off, and, without some kind of a miracle, the plant faced closing.

Despite advice from others that this airplane-building stunt could be a terrible waste of time and money, Budd forged ahead with plans that he had already laid out. His engineers had come across a small, two-wing amphibian airplane that was said to be very stable and easy to fly. The plane was the all-wood Savoia-Marchetti, which had been designed in Italy. It was a single-engine plane with four seats and had a 140 horsepower engine. The plane could take off from the water in a distance of 15 feet at a speed of 40 miles per hour. The cruising speed was 80 miles per hour, and the plane had a range of 375 miles.

The Budd Company purchased the plans, and the redesign and building of the world's first stainless steel airplane began in Philadelphia under the supervision of Col. Ragsdale. The structural members and the hull would be of stainless steel. The wings would be covered with fabric. The finished plane (Fig. 28) weighed 1750 pounds and was a little lighter than the wooden plane. They would call it the Pioneer. The thickness of the stainless steel was to be 0.006 inch, approximate-

Fig. 28 Budd Pioneer amphibian plane. Reprinted with permission of the Franklin Institute

ly the thickness of two sheets of paper. A one ton order for the steel would be placed with Allegheny Steel at Brackenridge, Pennsylvania.

The plane was test flown in the summer of 1932 and then demonstrated at airports in a number of American cities. The plane was then sent to New York to await shipment to Europe for flight demonstrations and to advertise the Budd shot weld process. Mr. Budd was particularly anxious to have the publicity in Europe.

Lou F. Argentin, the French Pilot. Lou Argentin arrived in New York aboard the *Île de France* in June 1929. He was 27 years old and had a new life before him. Lou had graduated from an engineering school and had immediately joined the French Air Force, where he was a pilot in a fighter squadron. After seven years in the Air Force, during which time he was awarded France's highest decorations, the Medaille Militaire and the Croix de Guerre, Argentin resigned.

He decided to see something of the United States before investigating the possibility of flying the mail in South America. While aboard the *Île de France*, he encountered a Mr. Eno Bossi, who was returning to the States with the rights to manufacture, under license, one of the top airplanes of the day, the Italian Savoia-Marchetti. After talking at some length and learning that Argentin was a pilot, Bossi invited him to come out to his plant at Port Richmond, Long Island, to see the Savoia-Marchetti.

After a month of sight-seeing in New York City, Argentin decided to visit Mr. Bossi. Argentin took the Savoia-Marchetti up for a brief flight and was intrigued with the idea of taking off at only 40 miles per hour and cruising at 80. It gave him time to enjoy the view. Argentin accepted Bossi's job offer and was soon demonstrating the plane to customers and flying the planes to their destinations after they were purchased. All went well until the stock market crash in October, and Bossi regretfully gave him his final pay check.

Argentin rather quickly found that there might be a job at the Chance Vought Aircraft Company in Hartford, Connecticut. He was interviewed at Chance Vought by a Dr. Michael Watter, who, interestingly enough, would later become head of the Research Division at the Budd Company. Argentin was hired and spent two years as a general inspector in their Experimental Department. While on business one day in New York City, Argentin decided to go out to Port Washington to see what was happening at Mr. Bossi's plant. As he pulled into the seaplane base, he couldn't believe his eyes. He saw a Savoia-Marchetti moored at the dock, but it was shining silver. The plane was the Budd stainless steel Pioneer.

Argentin found out that Bossi had sold the plans to the E.G. Budd Company, and they had redesigned and built the stainless steel plane, which was about to be shipped to Europe to make demonstration flights. Lou lost no time getting to the Budd Company in Philadelphia. When he was seated across from Mr. Budd's desk, Lou recited his credentials and his experience with the Savoia-Marchetti and said he would like to have the job of demonstrating the plane in Europe. Budd was overjoyed to find Lou Argentin and offered him a job on the spot.

Lou went back to Hartford and explained the extraordinary circumstances that led him to ask for his resignation from Chance Vought. Lou was soon on the *Île de France* with his American wife, on their way to Europe. The little plane was safely stowed in the hold below.

In Europe, Lou arranged to hold demonstration flights in France, Belgium, Germany, Sweden, and England. For Lou, it was more like a vacation than work. He flew over the English Channel once and over the Alps twice, once from France. On the return trip, Lou had to make a forced landing with the wheels retracted. He had the plane dismantled and hauled to a small stream, where he took off for the balance of the trip.

In England, when Lou was flying low along the coast, he misjudged his height, and the plane crashed into the water. A belly plate was broken, and the plane sank in four feet of water. It was there for 1½ or 2 days, after which it was taken out and repaired. There was no damage whatsoever to the plane as a result of its submersion in salt water.

The demonstrations abroad created quite a lot of interest, especially in France, where, of course, Lou felt very much at home and was warmly welcomed by his countrymen. The French government was particularly interested in stainless steel planes at that time. Lou explained the advantages of the Budd shot weld process, and plans were laid to set up a shop in Paris, where shot welding could be taught and licenses could be arranged. Many stainless steel aircraft floats were later made under license in France, and Lou brought home some very large orders for stainless steel oil coolers for French aircraft.

The highly successful trip was over, and it was beginning to look as if Mr. Budd's crazy idea was already paying off.

The Budd Company had no further use for the plane, and it was donated to the Franklin Institute in Philadelphia in 1935. It can be seen on display in front of the building, with a small plaque that reads: "Pioneer 1, World's First Stainless Steel Airplane. Built by the E.G. Budd Manufacturing Company, 1932." The metal is still rust free.

The World's First Stainless Steel Rubber-Tired Train

The Budd-Micheline Train. Turning back the clock to approximately 1928, legend has it that André Michelin, president of the Michelin et Cie, the French tire company, was on the night train from Marseilles to Paris and was kept awake half the night by the clickety-clack of the steel wheels against the steel rails. Michelin asked himself, "Why couldn't we put pneumatic tires on trains?" He had a wheel designed with a steel flange circling the tire that would keep the wheels on a railroad rail. Realizing that normal railroad cars were far too heavy to be supported on pneumatic rubber tires, he bought a small Citröen bus and had his special wheels mounted on its axles. He then took the bus for a ride on a section of railroad track. It worked just fine. The ride was smooth and quiet, except for the engine. Mr. Michelin had his invention on display for a while, sitting in front of a giant steam locomotive. The name Micheline was painted on the side of the bus, changing his name, Michelin, to the feminine form, as for a ship. A picture postcard at La Gare Deveau was made of the scene, which was entitled "Davide et Goliath."

In 1931, there was a surprising letter from France. M. Hauvette Michelin of Michelin et Cie, the tire company, wrote to invite Mr. Budd to come see the rubber-tired train. The two companies, of course, were well-known to each other with their joint ownership of the Budd-Michelin Truck Wheel Company that opened in 1924 in Detroit. Budd arrived at Paris in September 1931, where Michelin showed him the little Citröen bus with flanged wheels and rubber tires resting on a railroad track siding. They took a short ride so that Michelin could demonstrate the quietness and smoothness of the ride.

Michelin offered Budd the possibility of revitalizing the railcar industry in America, using Budd's high-strength stainless steel construction methods and Michelin's patented metal-flanged rubber tires. Michelin would exchange his patent for Budd's stainless steel know-how and shot weld patent. Budd accepted Michelin's proposal.

On returning to Philadelphia, Mr. Budd told his staff his idea. They would build an experimental Budd-Michelin self-propelled stainless steel railcar with rubber tires. The company had just completed delivery of a large stainless steel bus to the Chrysler Company, which then owned the Greyhound Bus Company. Making a railcar shouldn't be too much different, except that they would want even lighter construction.

Colonel Ragsdale and Albert Dean, a recently hired aeronautical engineer from Massachusetts Institute of Technology, comprised the team assigned to design and build what would be called the Budd-Micheline experimental train. Instead of attempting to emulate the traditional boxlike design of passenger cars, replacing the structural steel with high-strength stainless steel to reduce weight, Ragsdale and Dean decided to design it like a bridge, with diagonal braces, and use the stress analysis methods developed for aircraft construction.

Ragsdale, having completed the building of the Pioneer, was now in an excellent position to apply the acquired skills to the design of a light railcar. He had the added advantage of having an aeronautical engineer at his side. Ragsdale emphasized that lightweight construction required superior design, with more effective placement of materials and a considerable use of square and rectangular tubular members of very thin material. He said later that "the designer's effort to save weight may result in a seemingly complex structure but it should be remembered that every pound of stainless steel saved is also a savings of 45 cents." This had to do with the original cost, but the reduction in weight would also result in fuel savings for the life of the car.

The selection of the highest-strength stainless steel possible for the Budd-Michelin car was a half-hard, cold-rolled stainless steel alloy similar to what is now called type 301 stainless steel. It was the leanest and thereby the least expensive of the 18-8 category of stainless steels, permitting chromium and nickel contents as low as 16 and 6%, respectively. The designation "half-hard" for this steel meant that the minimum tensile strength of the steel would be 150,000 pounds per square inch, and the yield strength would be at least 110,000 pounds per square inch. The steel also would have a minimum elongation of 18%, as measured in two inches of length, which was ample for doing a reasonable amount of bending or forming of the metal into tubular shapes without fracturing.

The Budd-Michelin car, which was affectionately referred to as the "Green Goose," was ready for trials in January 1932. The car was 40 feet, 6½ inches long, including a 5 foot baggage compartment at the rear. It would seat 32 passengers and weigh just 13,500 pounds. Coaches normally had two axles at the front and rear, but the Budd-Michelin had three axles front and rear to reduce the weight on each tire. Therefore, the car had six tires on each side. The middle of the three rear axles was the driven axle.

The stainless steel sheet for the car was Allegheny Metal supplied by the Allegheny Steel Company of Brackenridge, Pennsylvania. A Junkers 85 horsepower engine at the front of the car was under a hood, giving the appearance of the front of an automobile.

Ragsdale and Dean next set about writing a patent application for a pneumatic-tired stainless steel car. It consisted of eight pages with five pages of diagrams. The patent, entitled "Railcar Body," was filed on April 27, 1932, and U.S. Patent 2,129,235 was granted on September 6, 1938. Some excerpts of the patent are as follows:

UNITED STATES PATENT OFFICE
Railcar Body
2,129,235
Earl W. J. Ragsdale, Norristown, Pa. and Albert G. Dean, Narberth, Pa.,
assignors to Edward G. Budd Manufacturing Co., Philadelphia, Pa.
Patent filing April 27, 1932
Patent granted September 6, 1938

"The invention relates particularly to vehicles of an order and size adopting them to travel upon standard railway tracks and to engage in traffic involving loads of the kind and degree encountered in connection with railway traffic as it exists today. However, the chief aim and end is the revolutionizing of the structural organization of the bodies of such vehicles to the end of achieving a maximum lightness of structure, of reducing the dead weight in such vehicles so low that the gross weight, dead load plus live load, is so enormously far reduced over that of the standard railway vehicle of today, through the use of a car body of the invention, that overlying invention of pneumatic tired railway vehicles may be practiced.

". . . The second aim of the invention is the achievement of a structure which may be constituted of electrically spot welded stainless steel in its largest part. The extremely great strength of properly treated stainless steel per unit weight as compared with the strengths of ordinary steels coupled with its eminent adaptability for electrical spot welding and its immunity from corrosion render it an ideal material for use in light weight construction, such as pneumatic tired railway vehicles."

The building of a second Green Goose, officially named the Lafayette, began immediately. It was shipped to Michelin in May 1932. Nothing is known of what happened to the Lafayette.

Meanwhile, Mr. Budd made it his job to sell the Budd-Michelin cars, inviting railroad executives to come to Philadelphia to take a ride in his rather odd new invention. There were just three buyers that decided to try rubber-tired stainless steel trains: the Reading, the Pennsylvania, and the Texas-Pacific Railroads.

The order from the Reading Company was received on May 5, 1932. It was for a considerably larger car than the Green Goose and would carry 47 passengers. It was equipped with a 125 horsepower Cummins Type H six-cylinder diesel engine at the front of the car under the floor. It was a single car, which the Reading Company officially named Rail Motor Car 65 (Fig. 29). It would be used as a commuter train on a ten-mile run between Hatboro and New Hope, Pennsylvania.

Fig. 29 The Budd-Michelin rubber-tired train, purchased by the Reading Company, was officially named Rail Motor Car 65. Source: Thum, *The Book of Stainless Steels,* 1935, p 429

As with many new models of automobiles, Reading Car 65 was plagued with problems from the start. On the second day of service, the fuel line became plugged, and the car had to be towed back to the shop at Budd. In the first six months, the car was out of service approximately half the time, and on April 8, 1933, a pheasant struck the front window, breaking the shatterproof glass. As if that were not enough, on June 18 at 9:30 in the morning, when the train was going along at 30 miles an hour, a motorcycle and side car struck the side of the Budd-Michelin at Bristol Street, north of Ivyland. The motorcycle was lodged under the rear truck, derailing all six wheels of the rear truck. Both occupants of the motorcycle were taken to the Abington Hospital. Car 65 was unharmed.

Despite the mishaps, Car 65 continued in service for 12 years and was retired from service on June 30, 1944.

Budd made only five of the experimental rubber-tired trains. They were found to be impractical, primarily because of tire puncture from rail slivers.

However, the building of these strange rubber-tired trains revolutionized the way passenger cars would be constructed and led to the development of a major industry, the construction of lightweight stainless steel railcars with steel wheels.

The Burlington Zephyr

Ralph Budd was born on an Iowa farm in 1879. Although no relation to Edward G. Budd, the two men would meet one day to enter into one of the most important business transactions of the 20th century. Ralph Budd was a brilliant young man who finished high school and college in just six years, having received a degree in civil engineering. He then went to work for the railroads and, at age 27, was chosen to work on the Panama Canal Project, where he was in charge of building a rail line across Panama's rough terrain.

In 1919, at the age of 40, Ralph Budd became the youngest executive of a railroad when he became president of the Great Northern Railroad. In January 1932, he moved again to become president of the prestigious Chicago, Burlington, & Quincy Railroad, a name that was usually shortened to just "the Burlington." The country was in the midst of the Great Depression, and the Burlington was in serious financial shape, as were all the railroads at the time. Freight traffic was down, and passenger traffic had dropped to the lowest since World War I, partially because more and more people were driving their cars instead of taking the train.

Ralph Budd was faced with a grim situation and spent most of his waking hours trying to think of something that might lure passengers back to the railroad. Meetings with the heads of all of the departments were held to brainstorm the problem. There were a lot of suggestions, including the lowering of fares, but nothing was suggested that really seemed practical. Budd kept asking the question, "How can we make riding a train more attractive?" Their trains were old and dirty from coal smoke. Heating was uncertain, and air conditioning did not exist.

Budd learned that a firm in Philadelphia was building lightweight passenger cars that were self-propelled and built of a new material called stainless steel. He found that the company was the E.G. Budd Company, and he mentally noted that the company was not one of America's four train builders: American Car and Foundry, the Pullman Company, Brill, or Bethlehem Steel.

Budd had been president of the company since January, with not much to show for his efforts. He arranged a visit to the E.G. Budd Company and was given a test ride on one of the Budd-Michelin trains. Ralph Budd was favorably impressed and said that they would hear from him. Because it took from the fall of 1932 until June 1933 for Ralph Budd to place an order, it seems fairly obvious that he was soliciting bids from the other car builders.

In the meantime, anecdote has it that Ralph Budd was relaxing one evening while reading *Ode to the West Wind*, in which the Greek god Zephyrus was the God of the West Wind. In a flash, Budd decided that he would call his train the Burlington Zephyr.

The Budds Sign a Contract. On June 17, 1933, Ralph Budd came to Philadelphia to discuss the contract. He told Edward Budd that he wanted a three-car train equipped with a Winton diesel-electric engine and steel wheels. The interior design would be left entirely to the E.G. Budd Company.

The contract was promptly signed and plans were laid. The lead designer was to be Albert Dean, who would modify and refine the design of the Budd-Michelin rubber-tired car. He used many ideas from aircraft design to enhance the streamlining. Trains had never been streamlined. Albert Dean completely enclosed the underside of the car to reduce drag, and he tested his design with a model in a wind tunnel at the Massachusetts Institute of Technology. It was later determined that the amount of wind resistance brought about by Dean's streamlining had reduced the drag by 47% at a speed of 95 miles per hour. The front of the train, which sloped down like a shovel for reduced wind resistance, was designed by an architect, John Harbeson.

There were no handles or other objects protruding, not even rivets. It would have smooth shot welds.

Walter Dean, who was Albert Dean's older brother, was in charge of designing the undercarriage of the cars, which included the heavy steel trucks that consisted of the wheel and axle assemblies that support the entire weight of the car. On conventional trains, each car is supported by two trucks, each with two axles and four wheels that are positioned at each end of the car. Walter Dean's idea, on the other hand, was to save weight by having the truck at each end of the center car share the load of the cars before and after. This permitted the train to be built with four trucks instead of six, for a weight savings of approximately six tons. He created what is called an articulated train, which permits the cars to turn on roller bearing pivots as the train rounds a curve. Two couplings were also eliminated.

The three-car train would be 196 feet long and weigh 104 tons, or approximately the weight of a single Pullman coach. It would be powered by a recently developed Winton 8-201A 660-horsepower, 2-cycle engine that would drive an electric generator. The generator current would be fed to electric traction motors in the train's front truck and would provide power for lighting, heating, and air conditioning. The unpainted sides of the train would be fluted stainless steel sheet to increase rigidity and enhance the appearance. It would be fast—maybe 100 miles per hour—and would quite often be called "The Silver Streak."

The train represented the largest use by far of stainless steel in any structure. Twenty-three tons of type 301 stainless steel was used in its construction. The stainless steel car body was built primarily with steel as thin as 0.012 inch, formed and spot welded to make hollow square and rectangular subsections (Fig. 30), which were then joined by spot welding. The hollow sections created the lightest possible structure, although requiring much more assembly time than using solid bars and plates. A lighter car body meant that the supporting wheel trucks could be lighter. The roofs were constructed of corrugated stainless steel 0.022 inch thick, or approximately the thickness of a stack of seven sheets of paper. Colonel Ragsdale said the roof would easily support a man walking on it.

Paul Phillipe Cret, head of the University of Pennsylvania School of Architecture and a noted interior designer, was hired to design the train's interior. He wanted the passenger compartments to reflect a state-of-the-art appearance in keeping with the exterior streamlining. The cars would look stark and unadorned when compared with the Pullman coaches of the day. There was to be indirect lighting, and the

Fig. 30 The Burlington Zephyr under construction. Courtesy of the Hagley Museum and Library

walls would be pastel colors of pale green, blue, and light brown. The plush armchairs in the observation car would be comfortable and covered with fine but sturdy fabric. There would be luxurious carpeting and curtains in shades of pearl gray. To preserve the clean lines of the interior, there were to be no overhead baggage racks. Instead, there would be a baggage compartment. Each passenger compartment would be individually heated and air conditioned. It would be the train of the future.

The entire period of designing, building, and decorating was one of intense excitement, with everyone working as if his job depended on it. The Zephyr (Fig. 31) was completed in just 9½ months, 2½ months ahead of schedule. The feat was all the more remarkable when it is considered that the equipment for the train had to be ordered from a hundred different suppliers. The cost of the train was approximately $260,000, which would be about $3,640,000 in the year 2008 when adjusted for inflation.

There was a final inspection with the first train. The first car held

Fig. 31 The Burlington Zephyr outside the factory. Courtesy of the Hagley Museum and Library

the diesel engine, an engineer's cab, a 500 gallon fuel tank, a 30 foot railway post office, and a space for baggage. In the second car, there was a larger baggage compartment, a buffet grill, and a 16 foot smoking section for 20 passengers. The third and last car, the observation car, had a 31 foot section with seats for 40 people and club chairs for 12. The train had a total of 72 seats and could carry 25 tons of baggage and express freight.

Two days after the train was completed, it was taken on a 25 mile test run from Philadelphia to Perkiomen Junction on a line that the Reading Company had cleared and inspected. The train ran perfectly and reached a speed of 104 miles per hour for a brief time.

The formal dedication came on April 18 at Philadelphia's Broad Street Station, directly across from the City Hall. At the appointed time of 10:00 a.m., the ceremony took place amidst a crowd of hundreds. Ralph Budd spoke over national radio, saying that "the Burlington Zephyr is a symbol of progress." On the following day, 24,000 people lined up to inspect the train. The Burlington Zephyr was then taken on a three-week tour to 30 cities in the East, during which time it was estimated that close to a million people went through the train.

However, Ralph Budd had a plan that would attract the attention of the entire nation. He was aware that the Chicago "Century of Progress" exposition would be opening its second season on May 25, 1934. One of the major attractions would be the "Wings of a Century Pageant," showing America's progress in transportation—from the Indians to the modern steam locomotive. Budd was going to display his Burlington Zephyr in that pageant, which opened Saturday evening, May 26, 1934.

Budd had been planning the big event for weeks. He would have the Zephyr leave Denver at dawn on May 26 and race 1015 miles nonstop to Chicago, pulling in at dusk, just in time for a grand entrance to the exposition at 9:00 p.m.

The Zephyr was moved to Denver on Thursday, April 24, and set on display at the Union Station. Newspapers reported that "a silver train has flashed into the silver state." Visitors went through the train that day and also on Friday, up to the time that the train was sent into the shop for inspection before the mad dash the next day. Celebrities and politicians got wind of the wild trip and were clamoring for tickets. However, all 85 seats were reserved, including 13 folding chairs set up in the baggage compartment, for officials of the Burlington Railroad, General Motors, Edward Budd, and reporters.

The Great Train Ride. Ralph Budd had made special preparations for the thousand-mile dash. All of Burlington's employees along the route had been notified to inspect the entire length of track and to post large signs at those places where the train should slow down for some reason. Railroad employees, police, and army veterans would man every one of the 1689 grade crossings to be sure that all road traffic would be stopped at the time the train would be passing. All passenger and freight trains would be sidelined. Guards would also be posted at each railroad station to make sure no one got onto the track.

Never before had such preparations been taken for a train trip, giving the train the clearest right-of-way a train would ever have.

At the shop, the mechanics found, to their horror, that there was a cracked armature bearing. No replacement was found in Denver, and calls went out to other railroads. A replacement was found in Omaha, and the bearing would be flown to Cheyenne, where it would be transferred to a chartered plane that would reach Denver at approximately midnight. Budd was going on NBC radio, and if the race was to be cancelled, now was the time to do it. However, Budd spoke into the microphone in a firm voice, saying, "Tomorrow at dawn we'll be on our way!" He invited the public to "come out and watch the train speed past."

By morning, the new bearing had been installed, and the train left the Union Terminal, breaking the timing tape at 7:04 a.m. In the driver's compartment were three men who would take their turns at the controls in two-hour shifts. These men were J.F. Weber, Burlington's superintendent of automotive equipment, J.S. Ford, an assistant master mechanic, and Ernie Kuehn, a Winton engineer. Three Burlington mechanics also rode in the cab. All of them would be scanning the road ahead to watch out for any problems.

The two Budds were seated side by side in the observation car, waiting for the show to begin. They had a great deal at stake on this maiden voyage.

The train was kept at 50 miles an hour for a while to break in the new bearing. It was gathering speed when a door slammed shut on the engine starter cable, causing burning rubber. The engineer shut off the engine and then could not restart it. There was a frantic search for a piece of wire with which to repair the starter cable. As the train slowed to 15 miles per hour, Ralph Budd burst into the engine cab and shouted, "Don't let her stop!" The men quickly repaired the wire, and the engine was restarted.

The train was now on its way, and the throttle was pushed to 90 miles per hour for much of the trip. She ran at 100 miles per hour along one 19 mile stretch and reached a top speed of 112.5 miles per hour for 3 miles.

It was a quiet ride, with only the clack of the rails and the brief clanging of the bells as the train passed the grade crossings. The telephone poles looked like picket fences. Ralph Budd had guessed that the trip might take approximately 15 hours, which would be possible if they could average a pretty fast clip of 67 miles per hour. The Burlington's crack train on the Denver-to-Chicago run, the Aristocrat, took almost 26 hours and stopped several times to take on water, coal, and new train crews.

About once an hour, one of the drivers would walk to the back of the train to hand Ralph Budd a note stating the miles covered and the present speed. Budd had a map so that he could trace the progress and pass the word along to others in the car.

The stations were crowded, and they could hear cheers as the train roared by. Reporters said that a million people came out to see the train speed past.

By the time the train reached Lincoln, Nebraska, which was 483 miles from Denver, the clock showed an elapsed time of six hours and seven minutes. One of the railroad officials announced that the train had just broken the nonstop record set by the Royal Scot in 1928.

News bulletins were dropped off at prearranged intervals and then telegraphed to radio stations and newsrooms. The train attracted bigger and bigger crowds. In town after town across rural Iowa and Illinois, fire sirens shrieked and church bells rang to give notice of the train's approach. Trucks and Model T's were clogging the roads next to the tracks.

It was 5:55 p.m. when the Zephyr streaked though Princeton, Illinois, passing the Burlington Aristocrat, which had left Denver 24 hours earlier. At 8:09 p.m., the Zephyr broke the tape at Chicago's Halstead Station, which timed the trip, and then continued on to the Chicago exposition grounds, fitting onto the 200 foot long stage with four feet to spare, and pulled to a stop amid a cheering crowd. Thousands of spectators poured out of their seats and up to the stage, where they could touch the train. Boats in the harbor blew their horns and whistles as if it were New Year's Eve.

Inside the train, there was cheering, laughing, hugging, and the shaking of hands. The two Budds congratulated each other over and over again. They had achieved something beyond their wildest dreams, and it was another of Edward Budd's stream of "world's firsts."

After order was restored, Ralph Budd stepped from the baggage room, leading the train's mascot, the mountain burro named Zeph, and presented him to the officials of the fair. "It was a sweet ride!" Budd exclaimed.

The trip, said Ralph Budd, demonstrated three things: "That the morale of the men and officers of the Burlington is proved by the way this run has been planned and carried out; second, the efficient condition of the railroad has been shown; third, that the train performs fully up to expectations."

Many records had been set. It was a world's record for the longest and fastest nonstop railroad run, having sped 1014.4 miles across ⅓ of the continent in just 13 hours, 4 minutes, and 58 seconds. The average speed was 77.61 miles per hour. The run was made in a little less than half the normal time. The diesel fuel used was just 418 gallons and cost an unbelievably low amount, $16.72, because diesel fuel was only four cents a gallon. The Burlington Aristocrat, on the other hand, had burned 85 tons of coal, costing $255 at $3 a ton. Putting it another way, the Zephyr got two miles to the gallon, and the fuel cost was 6.7% that of the steam train.

Newspapers across the land had screaming headlines. There hadn't been such good news since Lindbergh crossed the Atlantic:

The New York Times

Zephyr Makes World Record Run,
1015 Miles at Average of 78 Miles an Hour

On Denver-Chicago Dash, Streamlined Train Beats the Mark of Royal Scot, at Times Reaching 112 Miles an Hour Cuts the Usual running Time Nearly in Half

The Philadelphia Inquirer

Burlington Zephyr Eclipses Scot's Record in Non-stop Run,
Denver to Chicago; Averages 77.5 M.P.H.; Top of 112.5

Chicago, May 26 (A. P.) – Ending the first sunrise to sunset non-stop run in American railroad history, the Zephyr of the Burlington railroad swept into Chicago at 9:09 P. M. tonight, Philadelphia daylight time, at 96 miles an hour.

The trip changed forever the way passenger trains would be built, and it spelled the end of the smoke-belching steam engine.

The saga of the Burlington Zephyr was just beginning. Ralph Budd was now going to take his train on an extended tour of midwestern and western cities where the Burlington operated. The tour covered 222 cities, and two million people had a chance that summer to inspect the Zephyr. People, weary of the Depression, were delighted to see something new and exciting, and many couldn't wait to take a ride on it, which was exactly the point of Ralph Budd's tour. It couldn't have been better advertising for Edward Budd. In the coming years, he would have almost every railroad in the nation lining up to buy a train.

"Silver Streak." Other people saw the Zephyr in the movie "Silver Streak." The Zephyr was borrowed by Hollywood to be the star of a film showing the silver train streaking across the country to deliver an iron lung to a victim of polio just in time. The movie was not a great box office success, but it indicated the extent of a sort of "Zephyr mania" that was sweeping across the nation.

Finally, the great tour was over, and the Zephyr began regular train service on November 11, 1934, between Kansas City, Missouri, and Lincoln, Nebraska. Seats were sold out days in advance. At the end of November, a fourth car was ordered, and in the following year, the train made a profit of $90,000. At that rate, the train would pay for itself in just three years.

The train was renamed the Pioneer Zephyr on its second anniversary and remained in service for 25 years, having covered 3.2 million miles. It was then donated to the Chicago Museum of Science and Industry, where it remains on display inside the building.

The Pioneer Zephyr spawned a family of sister Zephyrs, each one a testimony to the Burlington Railroad's dedication to engineering excellence. It was the model for the passenger trains of competing railroads, although none of the other trains were made of stainless steel for many years, mainly by virtue of Budd's patented shot weld system.

The Budd Company would manufacture almost 11,000 stainless steel railcars over the next 50 years—enough cars to make a train stretching 175 miles, from Philadelphia to New Haven, Connecticut.

The Flying Yankee

> "Revolutionizing rail travel, The Flying Yankee remains one of the technological marvels of the modern world, and a testament to the quality of work undertaken by the welders and engineers who constructed it."
> American Welding Society, 2004

Having received its first order for a stainless steel train with a diesel-electric engine in May 1933, the Budd Company almost certainly embarked on an advertising campaign to attract other buyers, despite the fact that the stainless steel train was still on the drawing boards, untried and untested.

The advertising brochures could have included a description of the train, something as follows:

> BUDD COMPANY ANNOUNCES
> TRAIN OF THE FUTURE
> ATTRACT MORE PASSENGERS—DECREASE COSTS
>
> A three-car stainless steel train has been designed and patented by an aeronautical engineer with the following characteristics:
>
> **STRIKING APPEARANCE**—The silvery train is streamlined and strikingly beautiful. The use of stainless steel three times the strength of structural steel and the creation of a design to minimize weight will create a three-car train of about the weight of a single Pullman car. The train is articulated to enhance streamlining, reduce weight, and give it the appearance of being one long, seamless structure.

> **REDUCED OPERATING COSTS**—There is no locomotive. A compact 600 hp diesel-electric motor in the leading car powers the train. Fuel costs for oil rather than coal will be reduced by up to 90%. The job of fireman will be eliminated.
> **REDUCED MAINTENANCE COSTS**—The lower weight increases the life of wheels and track. Painting of the exterior will never be required.
> **COMFORT**—The train will have a specially designed interior with comfortable seating. There will be indirect lighting, controlled heating, and air conditioning. The train will be clean since smoke is virtually eliminated.
> **SPEED**—Speeds up to 90 mph will be possible so that traveling times can be reduced.

Although initially a somewhat more expensive purchase, at a price of $280,000, it would be noted that the very lightweight, 132-passenger train would pay for itself in a very short time because of greatly reduced operating and maintenance costs.

Other interesting features about the train include the fact that the motorman could sit comfortably, for the first time out of the weather, at the front of the lead car, commanding a clear view of the track ahead. Just behind the motorman's cab, the 600-horsepower diesel-electric engine will be installed to provide power to two traction motors to drive the front two axles. Space in the first car will be provided for U.S. mail and a baggage compartment.

In February 1934, as the Burlington Zephyr was nearing completion, the Budd Company received an order from the Boston & Maine and the Maine Central Railroads. The order was for virtually a carbon copy of the Zephyr, except for the interior decoration and the baggage racks that the designer, Paul Cret, did not use on the Zephyr but were to be used on the new train, to be called The Flying Yankee.

The Flying Yankee was completed in January 1935. It would be the only Budd train to operate on the Eastern coast until after the war. The following article appeared in *Railway News* describing its arrival:

> **The "Flying Yankee" Arrives at Boston**
> **Self-propelled, stainless steel articulated unit will meet fast schedules on the Boston & Maine and Maine Central between Boston, Portland, Me., and Bangor**
>
> "A self-propelled, three-section articulated high speed train of light-weight stainless steel construction, with seats for 140 per-

sons, has been completed for the Boston & Maine and the Maine Central by Edward G. Budd Manufacturing Company, Philadelphia, Pa., on an order received today from the Winton Division of General Motors Corporation.

"This train, known as the 'Flying Yankee,' and the first of its kind to be delivered to an eastern railway, moved into New England Saturday, February 9, 1935, arriving at Boston after a trip from Mechanicsville, N.Y., with a party of railway officers and representatives of the press.

"Before going into regular service between Boston, Portland and Bangor, the train will be placed on inspection at various points on the Boston & Maine and the Maine Central, plans for which cover several weeks. It will then go into daily service requiring a daily round trip of 730 miles over the lines of the two railroads."

Railway News

The train (Fig. 32) drew a crowd of 10,000 when it first visited Portland, Maine, and Nashua, New Hampshire, and a crowd of 20,000 reportedly showed up to marvel over it in Boston. Like the Zephyr, the second car in the Flying Yankee was the club coach, with accommodations for food service (Fig. 33), while the third car was an observation car with club chairs (Fig. 34).

Fig. 32 Flying Yankee in Nashua, New Hampshire, March 1935. Courtesy of Brian McCarthy, President, Flying Yankee Restoration Group, Inc.

Fig. 33 Coach cars. (a) Mark Twain Zephyr. Courtesy of the Hagley Museum and Library. (b) Flying Yankee. Courtesy of Brian McCarthy, President, Flying Yankee Restoration Group, Inc.

The Flying Yankee was the younger sister of the Burlington Zephyr and arrived just seven months after the Zephyr's dramatic beginning. The second stainless steel train, like the second voyager to America, isn't so well remembered. However, the Flying Yankee has risen from the ashes after 50 years of neglect on a deserted siding and will soon

Fig. 33 (continued) (b) Flying Yankee. Courtesy of Brian McCarthy, President, Flying Yankee Restoration Group, Inc.

be restored to new condition, riding the rails in the hills of New Hampshire, and crossing the nation on special tours.

In April 1935, the Flying Yankee (Boston & Maine #6000) was christened with a bottle of water from Lake Sebago, Maine, and began service on its rigorous route: Portland to Boston to Portland to Bangor to Boston and back to Portland, all in one day. Everyone in Maine was anxious to get a glimpse of the silvery speeding bullet. Barbara Graves, now of Kennett Square, Pennsylvania, and in her late 90s, told how she and her husband, Stuart, were thrilled to see the train as they sat on a bank near the track to watch the train speed by. They were on their honeymoon.

Passenger traffic dwindled over the years as more and more people used cars. The three-car Flying Yankee had become unprofitable, and the train was retired in 1957, after 22 years and 2.75 million miles. It had been successful beyond all expectations, including the fact that fuel consumption for the 106 ton train had averaged approximately 2 miles per gallon.

The train was donated by the Boston & Maine Railroad to the Edaville Railroad at South Carver, Massachusetts, where it languished for almost 40 years. Robert S. Morrell, the son of a railroad conductor,

Fig. 34 Observation cars. (a) Mark Twain Zephyr. Courtesy of the Hagley Museum and Library. (b) Flying Yankee. Courtesy of Brian McCarthy, President, Flying Yankee Restoration Group, Inc.

purchased the train in 1993 for approximately $200,000 and moved it to Glen, New Hampshire, to begin restoring it to operating condition. Morrell died before any restoration began.

Restoration of the Flying Yankee. In 1997, members of the Morrell family founded the Flying Yankee Restoration Group, Inc., and the train was moved to Claremont, New Hampshire, where restoration first commenced. The train was sold to the State of New Hampshire for one dollar, with the understanding that the train would be used for tourist rides and education.

In the meanwhile, members of the American Welding Society heard of the restoration and were reminded of the critical part that electric resistance welding had played in the very existence of such a train. It had been the Budd Company's Earl J.W. Ragsdale who invented a welding process that would neither soften the stainless steel structure nor destroy its corrosion resistance. The significance of his invention was demonstrated by the fact that no company, other than Budd, built stainless steel trains during the 20 year life of the patent. A group of the Welding Society members decided that the importance of the Ragsdale shot welding process should be duly recognized with an award ceremony.

On April 4, 2004, R. David Thomas, a past president of the American Welding Society, made a speech at Claremont, New Hampshire, to dignitaries of the State of New Hampshire, the owners of the Flying Yankee, and to fellow members of the Welding Society. He heaped praise upon "the 69-year-old 'shot-welded' stainless steel Flying Yankee three-car passenger train." Thomas then presented a plaque to Commissioner Carol A. Murray of the New Hampshire Department of Transportation. The inscription on the plaque was as follows:

The American Welding Society
Historical Welded Structure Award

is honorably bestowed upon the Flying Yankee in recognition of the advanced technology and innovative welding technique known as 'shot welding' that was developed by the Edward G. Budd Manufacturing Company to join stainless steel throughout this train. This process created a lighter and faster train thereby revolutionizing rail travel. The Flying Yankee remains one of the technological marvels of the modern world, and a testament to the quality of work undertaken by the welders and engineers who constructed it.

October 8, 2004

In 2005, the train was moved again, this time to the Hobo Railroad at Lincoln, New Hampshire, where restoration work would continue, including the major job of rebuilding the four 3-ton frames (the trucks) that hold the wheels and axles. The restoration was half completed in 2006. Brian McCarthy, who became president of the Restoration Group in 2007, says he hopes the train will be completed by 2012, but it all depends on the raising of funds. The total job may cost approximately $6 million. The Restoration Group, as part of its fundraising efforts, has started selling naming rights to the 132 seats and the 3 cars for prices ranging from $2500 for seats to $500,000 for naming a car. In addition to a brass nameplate, the donors will be entitled to a certain number of train rides each year.

When restoration is finished, the Flying Yankee will run on short excursions from Lincoln, New Hampshire, every summer. The rest of the year it will be based in Concord, New Hampshire, and used as a rolling classroom for lessons on history, technology, and the advantages of railroads as the most economical form of transportation and the least damaging to the environment.

The Flying Yankee will be the only train of its type (one of the five articulated stainless steel trains built by the Budd Company) in the world to be in operating condition.

The Mark Twain Zephyr

The Mark Twain Zephyr was christened on October 25, 1935, at Hannibal, Missouri, which was Samuel Clemens' birthplace. The four-car, self-propelled, streamlined, stainless steel train was the fifth and last articulated train to be built by the Edward G. Budd Manufacturing Company at Philadelphia. The Twin Cities' Zephyrs had already been delivered to serve the Burlington's twin cities, St. Paul and Minneapolis in Minnesota.

The train was the third of what have been called "the Burlington's gleaming symphony of 18-8 stainless steel, the legendary Zephyrs." The first Budd train, the Burlington Zephyr, had been delivered to the Chicago, Burlington, & Quincy (CB&Q) Railroad in May 1934 and soon raised patronage to the point where there was a long waiting list on the route from Kansas City to Lincoln, Nebraska. Despite the fact that the country was in the very depths of the Great Depression, Ralph Budd's bet on investing in a streamlined train was, in fact, paying off handsomely in passenger revenues.

Burlington's management was hopeful that the Mark Twain Zephyr would do equally well on the 220 mile run from St. Louis to Burlington, Iowa. The advertising department had selected the train's name early on, because they were mindful that the town of Hannibal was halfway between the terminals. Furthermore, as luck would have it, Hannibal was celebrating the centennial celebration of the famous author, Samuel Langhorne Clemens—the immortal Mark Twain. The train would be christened the Mark Twain Zephyr. Zephyr had been selected for the first train by Ralph Budd, president of the CB&Q, because he had found that Zephyrus was God of the West Wind.

The Mark Twain would become the first Zephyr to have names for each car—Tom Sawyer, Huckleberry Finn, and Becky Thatcher. The day of the christening was one of the most memorable in the town of Hannibal. The Mark Twain arrived at midmorning, carrying railroad executives, newspapermen, and guests, among whom was the beaming, white-haired builder of the train, Edward G. Budd. The Hannibal Chamber of Commerce hosted a luncheon at the festooned Mark Twain Hotel. The menu included such dishes as "Life on the Mississippi Catfish" and "Aunt Polly's New Apple Pie with Cheese." There was music by the Bates Ensemble. It was the happiest of occasions. Special guests included Ossip Gabrilowitch, who was the conductor of the renowned Detroit Symphony Orchestra, and his wife, Clara Clemens Gabrilowitch, who was the daughter of Samuel Clemens, and their daughter, Miss Nina Gabrilowitch.

At 2:00 p.m., October 25, a coast-to-coast radio broadcast began at the railroad station, as did the dedication ceremonies. The honors of the christening ceremony fell to Miss Gabrilowitch.

The first regular service of the Mark Twain Zephyr began on October 27, 1935. The train was so well received that passenger ridership on the line increased dramatically. The Mark Twain was in service for 28 years before it was retired in May 1963.

Transit and Trucking

Urban Mass Transit. In 1934, Budd entered urban mass transit business with the completion of a subway or elevated train for the Brooklyn-Manhattan Transit Company in New York City. The train, numbered 1029, was put in service in August. It was 168 feet long and had a seating capacity of 640. It was called the Zephyr, like its diesel-electric brothers.

Highway Trailers. Budd also set up a Highway Trailer Division in approximately1934 and started building stainless steel trailer bodies for the booming trucking industry. The company performed tests that showed Budd trailers to be stronger and lighter than the average trailers on the market, and they could carry a greater payload.

The trailer business grew steadily. In 1940, Fruehauf, the large trucking company, became the Budd distributor and ordered 10,000 trailer bodies, which were shipped disassembled.

Train Sales. In 1936, Budd produced its first standard (nonarticulated), lightweight, streamlined train for the Atchison, Topeka, & Santa Fe (AT&SF) Railroad. (The articulated train design was abandoned because other cars could not be connected to one of those trains.) Santa Fe Coach 3030 soon led to an order for nine more coaches. Then came Budd's first really big train order: an order for 104 cars from AT&SF for delivery in 1937. Now, Budd had to set up a production line and could bring back more men of those previously laid off.

Other orders came in, so that by 1941, when train building ceased because of World War II, the Budd Company had built and sold 47 complete trains to 14 railroads.

However, railroad executives and engineers were generally not convinced that a railcar weighing only one-third that of a regular steel car could be as strong as the conventional steel cars. This led to the building in 1939 of a quarter-million dollar testing facility that could test an entire car in compression lengthwise. There had been no such facility in the United States (see the section "The Big Squeeze" in this chapter).

Train building continued at the Budd Company until 1987, including Metroliners, trains for Amtrak, and hundreds of subway and elevated cars for Philadelphia, New York, and Chicago. Over 10,000 cars were built in all. Appendix 2, "A Stainless Steel Timeline," lists many of the major purchases, while purchases made immediately after World War II are listed in the section "The Postwar Years" in this chapter.

The War Years

When the United States entered World War II in December 1941, many changes were to be made at the Budd Company. Work on civilian products had to be phased out as rapidly as possible to make way for military orders. During World War I, Mr. Budd once said, "I hate war and I hate war work, but I will never fail to serve my country."

Train building came to a halt, and army trucks and ambulances soon replaced the stamping of automobile bodies. There was a division that did Navy work and another group that built fragmentation bombs. Work in one building was so secret that even the workers did not know what they were making.

The "H" building had no other identification, and no one in the building could say what they were making, because they didn't know. They were brazing cones and nozzles to pieces of pipe approximately four feet long and three inches in diameter. Eventually, there was an article in *Life Magazine* describing a miraculous projectile that a soldier could fire from his shoulder to stop a tank at a distance of 100 yards or so. It was called the bazooka. General Eisenhower said after the war that "the bazooka was one of the four most important weapons of the war."

The World's First Stainless Steel Cargo Plane

In 1940, the Budd Aviation Division became active. The group consisted of Albert Dean, the aeronautical engineer who had designed the Burlington Zephyr, his brother, Walter Dean, Col. Ragsdale, and Lou Argentin, the French engineer and pilot. Dr. Michael Watter, an aeronautical engineer recently hired from Chance Vought Aircraft, was the sixth member of the team. These men were looking into the possibility of building a cargo plane for the Navy, which had released its specifications for a twin-engine cargo carrier and troop transport that was similar to the Douglas DC-3 passenger plane designed in 1937. The major differences from the passenger plane were a stronger floor and a large door on one side for the loading and unloading of cargo.

The Budd group came up with some interesting and important innovations, the first of which was that they would build the plane of stainless steel. The next was that the flight crew of two pilots and a navigator would be moved above the cargo space at the nose of the plane, so that the cargo space could extend the entire length of the plane. The crew would reach the flight deck by a set of stairs. The cargo space would then be 25 feet long by 8 feet wide by 8 feet high. Another idea was to have the tail sweep up so that a ramp could be let down at the rear of the plane, allowing an ambulance or a jeep to be driven right into the plane. Instead of a wheel under the tail, there would be a nose wheel and two wheels toward the middle of the plane.

Sketches of the Budd design were taken to a Navy office in Washington. The officers were quite impressed with what appeared to be some important advantages of the Budd design. It was somewhat harder to explain that a plane which was to be made entirely out of steel could actually fly. They pointed out that they were the world's experts in building lightweight stainless steel railcars and had built 600 of them. The Budd people explained that the unusually high strength of stainless steel made it possible to use very thin material to create a very light structure. It was also mentioned, while unrolling a large photograph to show the Pioneer plane in flight, that Budd had already demonstrated, back in 1931, its ability to build and fly a stainless steel airplane, albeit on a very small scale. The Budd Pioneer weighed just 1750 pounds.

The Budd group thought they had been well received, but realistically, the odds of getting a contract seemed remote. They were not an aircraft company; building airplanes made of stainless steel was a questionable concept, and the Budd design, although interesting, was quite radical. They couldn't even prove that the plane would fly, and their engineering staff included only two aeronautical engineers. On the plus side, however, their airplane would require the use of no aluminum, which was in exceedingly short supply.

The Budd people were dumbfounded to receive an order two weeks later, in August 1942, from the Navy for 200 airplanes, along with notification that the U.S. Government would build a factory for their production. The Budd price of $1 million per plane was accepted.

Hardly was the ink dry on the Navy contract when a contract was received from the Army, this time for 400 planes. Their total orders for the planes were now $600 million.

A 540 acre tract was found in Bustleton, a sparsely populated section of northeast Philadelphia. Red Lion Road ran along the property, and the aircraft factory was to be built there. It would be called the Red Lion Plant. There was sufficient land on which to build an airfield and provide parking space until the planes could be flown to their destinations.

While the immense factory, with an assembly area of 13 acres, was being built, the Budd engineers were creating the final design of what would be called the RB-1 Conestoga and making drawings of every component of the stainless steel structure and the steel undercarriage, which was not retractable in those days. The other items that had to be provided for included the flying instruments, Pratt & Whitney engines, Hamilton propellers, and wheels and tires. A bill of materials had to be made for all of the sizes of stainless steel sheet that would be ordered from the Allegheny Ludlum Steel Company. Approximately two tons of stainless steel would be needed for each plane. Lou Argentin, the French pilot who had flown Budd's first plane, would be the supervisor of final assembly. Frank Piasecki, the young future inventor of the Piasecki tandem rotor helicopter, would design the propulsion system. The factory was built in record time, and an assembly line was set up. The author, an employee of the Budd Company at the time, was visiting the aircraft plant and saw with wonder the size and shining appearance of the first RB-1 when it was almost completed.

On October 18, 1943, the first Budd RB-1 Conestoga (Fig. 35) was completed and had undergone an extensive series of static load tests, using equipment designed and built by Henry H. Weisheit, now of Lansdale, Pennsylvania. Flight and maintenance manuals had been prepared by Karl G. Reed, a cousin of the Dean brothers, who is now of Kennett Square, Pennsylvania.

The test flight was made on October 27, 1943. After circling the field for 35 minutes and being put through all Civil Aeronautics Association tests, the pilot landed and enthusiastically told the assembled designers, engineers, and builders, "You got an airplane."

Fig. 35 Budd RB-1 Conestoga cargo plane. Courtesy of the Hagley Museum and Library

Seventeen planes were delivered and accepted by the U.S. Navy. After the 25th plane was built, the contract was cancelled "because of strategic and logistic demands." The 100 foot wingspan "flying box cars" were no more.

One of the planes was sold to the Tucker Motor Company to fly its demonstration automobile around the country.

Seventeen of the planes were purchased for $30,000 each by the Flying Tigers, the first air freight company in the United States. The planes constituted the company's entire fleet for several years.

The rear-loading ramp feature, which was a Budd innovation, was adopted as standard for all future military cargo planes. It had been another "first" for the Budd Company, the first to build a stainless steel cargo plane.

The Postwar Years

The Budd Company had served well in World War II, with its 20,000 employees turning out army truck and ambulance bodies and fragmentation bombs. The building of the giant stainless steel cargo planes was another big effort, albeit too late to be put into service. Hundreds of employees went into the service, and an equal number of women took their places in the factory.

Now was the time to get back to civilian work as quickly as possible. Budd expected a big need for automobile bodies as the automobile industry got back to building cars, as well as a large demand for stainless steel trains, a business that had hardly gotten off to a good start before the war. Budd had sold just 600 railcars since they entered that business in 1934.

The Red Lion Plant was bought from the government right after the war. It had been built by the government for Budd's production of the giant cargo planes in north Philadelphia, approximately five miles from the Budd factory on Hunting Park Avenue in the Nicetown section of Philadelphia. It would now be used for auto body and railcar production.

It was a 13 acre plant, approximately the length of two football fields with a 100 foot ceiling. There were four 10 ton traveling overhead cranes. There would be enough space in the building for 80 railcars. The attached airfield would be used for railcars awaiting shipment. There was already a railroad siding.

The Railcar Division at the Red Lion Plant was set up as soon as possible. Letters were sent to the American railroads announcing that

Budd was back in business and enumerating the advantages of stainless steel cars with light weight, which saved on fuel costs and never needed painting.

The rail companies, with passenger cars that were worn out, were eager to buy. The CB&Q, Denver Rio Grande, and Western Pacific Railroads placed the first order. It was a substantial order for 66 cars. The CB&Q already had three Zephyr trains: the Pioneer I, which had first been called the Burlington Zephyr, the Twin Cities' Zephyrs, and the Mark Twain Zephyr.

From 1945 through 1950, the following orders were received for the Budd railcars:

Summary of early postwar train sales

Year	Number of cars	Railroads
1945–1946	27	Atchison, Topeka, & Santa Fe
1946–1947	16	Air Coast Line
1947	161	New York Central (127); Seaboard Air Line (16); Pennsylvania (10); Richmond, Fredericksburg, & Potomac (8)
1948	269	Florida East Coast (14); Missouri & Pacific (6); Minneapolis & St. Louis (6); Atchison, Topeka, & Santa Fe (16); Chicago, Burlington, & Quincy Twin Cities' Zephyr (19); Chesapeake & Ohio (45); Chicago, Burlington, & Quincy California Zephyr (27); Denver & Rio Grande California Zephyr (24); New York Central (112)
1949	207	Pennsylvania (78); Seaboard Air Line (6); Missouri Pacific (6); Delaware Lackawanna & Western (4); Southern (33); New York City Transit (10 subway cars); Central of Brazil (60); Louisville & Nashville (4); Chicago, Burlington, & Quincy (2); Great Northern (4)
1950	263	Atchison, Topeka, & Santa Fe (27 sleepers); Union Pacific (50 sleepers); Southern Pacific (30 sleepers); Chicago, Burlington, & Quincy (30 galleries); Southern Pacific (48); Soracabana (51); Western Pacific (1 Rural Diesel Car, or RDC); New York Central (4 RDCs); Chicago & Northwestern (3 RDCs); Pennsylvania-Reading Seashore Line (6 RDCs); Seashore Line (6 RDCs); Western Pacific (1 RDC); Baltimore & Ohio (2 RDCs); New York, Susquehanna, & Western (4 RDCs)

In 1945, the New York Central ordered 112 cars, followed by 127 additional cars in 1947. The Pennsylvania Railroad ordered 64 cars in 1951 for their flagship Boston-Washington train, The Senator, and the New York-to-Washington train, The Congressional.

In 1954, the Canadian Pacific Railroad placed the largest order to that date for standard passenger cars in Budd's history: an order for 173 cars, worth $40 million (approximately $290 million in 2005 dollars). The cars would be used on the Canadian Pacific's 18 prestigious transcontinental trains.

The Big Squeeze. Although the stainless steel trains were striking in appearance, very fast, and only approximately one-third the weight of an ordinary Pullman coach, there were many in the railway industry who believed such a light structure was flimsy and would crumple in any kind of collision. The Budd engineers, however, claimed that the Budd cars were just as strong, or even stronger, than the average passenger coach. Their calculations demonstrated this. However, many remembered the unsinkable *Titanic* and remained unconvinced.

The sale of the Budd trains had not come up to expectations, and it was decided that steps should be taken to demonstrate and measure the compression strength of a railcar, squeezing it lengthwise from end to end.

What was said to be the world's largest hydraulic, horizontal compression testing machine was ordered in 1939. It was to have a capacity of 2 million pounds, despite the fact that the Association of American Railroads (AAR) required passenger cars to pass an axial load test of only 800,000 pounds.

A test house was built with a track leading into it, and the big machine was installed on the Budd property at Hunting Park Avenue. In the meantime, plans were made to make the most of the demonstration. Not only would they make a compression test, but they would gather a great deal of information about the strength of the stainless steel structure and how it might vary from one section to another. The plan was to use the recently invented electric strain gages, of which Frank Tatnall of Philadelphia was one of the inventors. The paper-thin gages, which looked like adhesive bandages with wires leading from them, were cemented to a skeleton car frame at dozens of places, with the lead wires connecting to a central recording device. The gages, which consisted of metal foil one-tenth the thickness of a human hair enclosed in paper for electrical insulation, could detect the slightest deformation of the metal to which they were attached.

The gages would serve several purposes. They would show which sections of the structure were the strongest and which were the weak-

COLONEL RAGSDALE ON THE BUDD COMPANY PHILOSOPHY

Colonel R.J.W. Ragsdale was a graduate of the Massachusetts Institute of Technology, Chief Engineer of the Budd Railcar Division, inventor of the shot weld system, and a co-inventor of a lightweight stainless steel railcar. In 1941, Ragsdale wrote the definitive, 140-page illustrated book *Testing of Railway Passenger Cars and the Materials of Construction*. The book was published by the Edward G. Budd Manufacturing Company for special distribution, with only 100 copies printed.

Ragsdale explained why his company invested over one-quarter million dollars in full-car testing equipment:

> "The Budd Company has made this investment in equipment and the engineering talent to operate it for the purpose of proving its product to itself as well as to those railroad customers who have entrusted to the Budd Company their responsibility for safeguarding the lives of their customers."

He described the philosophy of the railroads at the time:

> "It is still common belief that weight represents strength. American railroad men have translated this into their equipment. They have built heavy and they have built for safety. Be it said to their credit that safety has been placed above cost, for it costs money to haul additional weight."

He described Budd's entry into the railcar business:

> "The Budd Company entered the business offering new structural development with stainless steel which was several times stronger than the steel previously used in car construction."

Ragsdale showed how full-car testing confirmed Budd's claims of high strength for their car, exceeding the Association of American Railroad's specified compressive strength by as much as 70%.

Ragsdale concluded his treatise by comparing the properties of stainless steel with the other high-strength, lightweight competitive materials of the day, notably, duralumin, an aluminum alloy, and Corten, a high-strength, low-alloy steel manufactured by the U.S. Steel Company. He showed the superiority of the properties of stainless steel, while at the same time noting the much higher cost of stainless steel.

est, so that future cars could be built with design modifications to reduce the amount of material at the stronger areas and increase the material, if needed, in any of the weaker areas.

The compression test would provide some other vital information. In addition to demonstrating, it was hoped, that the design of the car was adequate to provide sufficient strength in compression, it would also serve as a proof test of all the spot welds, because every part of the structure was interconnected.

When the spot welds were made, they left a smooth, slightly tarnished, very small depression in the surface, but there was no way of knowing that a given weld was strong and soundly made, unlike electric arc welds that leave a raised bead of metal. The answer to this seemingly unsolvable enigma was provided by Col. Ragsdale in the form of one of the best quality-control systems ever devised, which would give assurance that the welds were satisfactory.

In accordance with this plan, it is believed that test welds were made at the beginning of each shift for each welding machine, and if the twist test was satisfactory, the production welding could proceed. Ragsdale's requirement for the twist test was that the sample weld should withstand a twist of 60 to 90° without breaking. It had been determined that the twist test was an acceptable substitute for the pull test in assessing the quality of a weld. The test specimens were twisted by hand.

Each of the welding machines had an alarm bell that rang if a weld was not made according to the settings on the machine.

On the day of the big test, the load was increased slowly, so that the engineers had time to check the readings from the strain gages. The load reached 1,340,000 pounds, almost 70% more than that required by the AAR. The sturdiness of the Budd-designed car had been clearly demonstrated, and the Budd engineers could breathe a sigh of relief. The test car was not damaged and was used as an exhibit.

Slumbercoaches. Budd Slumbercoaches were developed in 1956. These cars brought a new concept to railroad coach travel by providing private sleeping accommodations at coach fares, plus a small room

A TRAIN COLLISION

"The superiority of stainless steel in this regard was illustrated in Australia in 1975. A B-Class diesel locomotive drawing carbon steel railcars was coming down a hill at Frankton, southwest of Melbourne. The brakes failed and the locomotive ran into the back of a stationary suburban train sitting in a station at the bottom of the hill. (The stationary train consisted of stainless steel railcars manufactured by Hitachi.) The rear stainless steel railcar, which was the first to be hit, ended up sitting on top of the B-Class locomotive, which lifted it up as it ploughed underneath. No one was killed and passengers escaped with minor injuries."

Nickel Magazine, December 1999

charge. The Baltimore & Ohio Railroad bought three coaches for service between Baltimore-Washington and St. Louis, two for their Columbian between Baltimore-Washington and Chicago, and four for the Burlington Denver Zephyr operating between Denver and Chicago. Each room had window seats by day and a full-sized bed at night, with private toilet facilities in a private room with individually controlled air conditioning and heating.

The Budd Rural Diesel Car (RDC). From 1959 to 1966, the Budd Company produced a self-propelled, diesel-hydraulic passenger car. These cars became very popular in rural areas with low passenger traffic density and for commuter service. The RDCs were much less expensive to operate than a traditional car with a locomotive. Almost 400 of the RDCs were built.

The RDC was essentially Budd's 85 foot railcar. The RDC-1 was an all-passenger coach that carried 90 passengers. The RDC-3, with a railway post office and baggage compartment, carried 49 passengers. Two RDCs could be coupled together to make a two-car train.

The Boston & Maine Railroad owned most of the RDCs, which served to replace the Flying Yankee that was retired in 1957. They were also very popular on the New Haven Railroad, with trains going to Cape Cod. Other cars were purchased by the New York Central, Reading Railroad, Pennsylvania-Reading Seashore Lines, Baltimore & Ohio, Jersey Central, and Canadian Pacific.

Three RDCs were shipped to Australia, and five others were built under license in Australia.

In 1966, as a publicity stunt, the New York Central fitted two jet engines on the top of one of its RDCs and set the U.S. speed record of 184 miles per hour between Butler, Ind., and Stryker, Ohio.

In many cases, the RDC cars were the last rural passenger trains to run in North America, and only a few are still in service. VIA Rail, Canada, still uses RDCs on Vancouver Island and in northern Ontario.

The RDC Roger Williams is at the Railroad Museum in Danbury, Connecticut.

Urban Transit Vehicles. In 1960, 270 subway cars were ordered for the Philadelphia Market-Frankford Elevated Line. In 1963, the New York City Transit Authority placed an order for 600 subway cars, which were delivered between 1964 and 1965.

In 1969, the Port Authority Transit Corporation ordered 75 rapid transit cars for the Lindenwold Hi Speed Line that connected southern New Jersey communities with Philadelphia. In 1980, there was a $150 million contract from Amtrak for 150 Metroliner cars.

Chicago became the largest owner of Budd cars, with 750 ordered for their transportation system between 1969 and 1984. The 1984 shipment was the last from the Budd Company. The company had been bought by Tyssen AG, Germany, in 1978. The Railcar and Auto Body Divisions were closed. Ridership had fallen, while personal auto travel and air travel increased.

Dulles Passenger Transporter. In 1962, the Dulles International Airport opened near Washington, D.C. A large passenger-carrying vehicle was built to move passengers from the terminal to planes waiting on the field. The transporter was built as a joint venture between the Chrysler Corporation and the Budd Company, with Chrysler providing the lower structure, wheels, and power and Budd providing the stainless steel air-conditioned lounge.

It was the largest passenger vehicle ever built to be operated on rubber tires. The mobile lounge was approximately 54 feet long, 16 feet wide, and 17 feet high and weighed approximately 76,000 pounds. The passenger compartment accommodated 90 seated and 17 standing passengers.

The lounge eliminated long walking distances for passengers, and the elimination of taxiing saved fuel. Eighteen more of the mobile lounges were produced by Budd and Chrysler.

A Review of the Budd Era

Over a period of 54 years, the Budd Company was one of America's largest consumers of stainless steel. In all, Budd produced approximately 10,000 railcars, using about 80,000 tons of stainless steel—enough cars to make a train stretching from Philadelphia to New Haven, Conn.

The Great Depression struck the Edward G. Budd Manufacturing Company without warning in November 1929, with 10,000 automobile workers facing layoffs unless, by some miracle, a new kind of work could be found. None of Budd's staff had any practical ideas. However, Mr. Budd decided to build a much stronger stainless steel structure than had ever been built in the 18 years since the alloy had been developed in Germany. Using his own money, he put his staff to work building a small stainless steel airplane. It flew and was sent on exhibition flights across America and Europe.

It seems reasonable to believe that the flight of the Budd plane in France attracted the attention of the Michelin tire company, because shortly afterward, there was an inquiry from that company about the possibility of building a very, very light railcar that could ride on Michelin tires. A light railcar was designed and patented by Budd's aeronautical engineer, and a self-propelled railcar was built for Michelin with six rubber-tired wheels on each side.

Budd built and sold three similar cars to three American railroads. Then, in 1932, the CB&Q Railroad ordered a three-car self-propelled stainless steel train that would have steel wheels and a diesel engine. Budd had revolutionized the way passenger cars were built, and orders started to come in, slowly at first but then by the hundreds of cars later on.

Budd, more than any other industrialist, made the American public aware of stainless steel. Americans were riding the shiny, streamlined stainless steel cars of 70 railroads, the Metroliners, and the Pennsylvania's Senator and the Congressional. Millions of commuters in New York City, Philadelphia, Baltimore, Miami, and Chicago were riding stainless steel subways and elevated trains. The Budd Company placed many advertisements with an artist's renderings of Budd trains going through canyons, mountains, crossing the plains, and traveling past New England towns.

In 1960, the Allegheny Ludlum Steel Corporation published *Strength of Stainless Steel Structural Members as a Function of Design*, a book authored by Dr. Michael Watter, Director of Research at the Budd Company, and Rush A. Lincoln, an engineer at Allegheny Ludlum. They dedicated the book to Mr. Budd, who had died in 1946, and included a letter that Mr. Budd had written to Mr. Batcheller, an official at Allegheny Ludlum:

DEDICATION to EDWARD G. BUDD

Whose vision and faith in stainless steel as a structural material instigated the research on which this book is based. Mr. Budd's plans and enthusiasm are shown by excerpts from a letter written a few days before his death and remain an inspiration to all who produce and work in stainless steel.

November 30, 1946

Dear Mr. Batcheller:

I want you to go to the Red Lion Plant . . . you would see us making side frames for railway cars in quantities . . . (they) would have looked to you just like a bridge member . . . this is the cheapest way to make a bridge . . . we can double the capacity of the bridge by changing the structure of the floor.

I am enclosing a picture of the airplane we built out of this metal in 1931. It has flown all over the Alps and over the Channel a number of times. It has flown all over the United States. It has stood outdoors in front of the Franklin Institute for years, where it now is. The sheathing of the plane is of steel 0.006 inch thick. This has no protective coating whatever and has been exposed to salt water, to storms, and to the action of the atmosphere for 15 years. The airplane industry will use stainless steel because it has strength and resistance to corrosion.

There are many machine parts where weight is an important factor. Gradually, as the arts refine our manufacturing processes, the manufacturers will use stainless steel in these parts. Cages for elevators would give greater strength and save a double portion of the weight, and on fast running elevators, weight is the important factor.

You are the manufacturers who have led in this stainless steel business. We are the fabricators who have given it the most support, and now lay before you a brilliant future. It is a beautiful metal and that is a factor in its favor. We use it because it is a strong, ductile, and practically indestructible metal. I don't believe there is any subject so vital to you as is contained in this letter.

Edward G. Budd, President
The Budd Company

What is left? There is the first stainless steel airplane, The Pioneer I, which still sits proudly in front of the Franklin Institute in Philadelphia (Fig. 36). There is the first train, the three-car Pioneer Zephyr built in 1934, which has been restored and is on display at the Chicago Museum of Science and Industry. There are 18 transcontinental trains in Canada that were restored in 1994 and could still be in service until 2034. And then there is the Flying Yankee, Budd's second train, which is going to be completely restored by 2013, if things go according to schedule. It is planned to have the train on display and for train rides. It is just possible that the Flying Yankee may still be running by the year 2100!

Fig. 36 The first stainless steel aircraft built by the Budd Manufacturing Company on display in front of the Franklin Institute Museum in Philadelphia, Pennsylvania. Courtesy of Craig Clauser

CHAPTER 9

The Gateway Arch

"A landmark of our time"
Eero Saarinen

WHEN LEWIS AND CLARK returned to St. Louis in 1806 after their three-year expedition to explore the territory of Thomas Jefferson's 1803 Louisiana Purchase, the city became known as the "Gateway to the West" for the many settlers, mountain men, and adventurers going into the western frontier.

In 1933, St. Louis Mayor Bernard Dickman assembled a group of businessmen who formed the Jefferson Expansion Memorial Association to honor the thousands of pioneers who had gone through St. Louis to settle the lands west of the Mississippi River. A Joint Resolution of Congress sanctioned the plan to build a memorial park in St. Louis. With a $7 million bond issue and $7 million in Works Progress Administration funding, the acquisition of land and the clearing of 40 acres along the Mississippi River began.

Little else happened until 1947, when the noted Finnish-American architect Eero Saarinen won a $22,500 architectural competition for his design of a memorial structure. St. Louis, known as the "Gateway to the West," led Saarinen to design what would be the tallest monument in the world, except for the Eiffel Tower. Saarinen designed an arch that would be 590 feet tall, almost 40 feet taller than the Washington Monument. It would be built of steel and concrete and faced with 426 stainless steel plates. The stainless steel would weigh 886 tons, more stainless steel than had ever been used on any project. The first stainless steel to be used on any building was the tower of the Chrysler Building, which was covered with 27 tons of the metal.

For some reason, Saarinen added 40 feet to the height, bringing it to 630 feet before construction started, and the width at the base between the legs of the arch would be 630 feet. The shape (Fig. 37) is that of an inverted catenary. (A catenary is the shape of a chain that dangles from two points at the same level.) No structure of this shape had ever been built. Perhaps Saarinen's idea stemmed from the fact that the famous Ead's Bridge, built in 1874, was still standing and would be a stone's throw from the location of his arch. The Ead's Bridge, of "true steel," was of a daring design with a 520 foot long arch, the longest arch of any bridge at the time.

Saarinen later said, "The major concern here was to create a monument which would have lasting significance and would be a landmark of our time. An absolutely simple shape—such as the Egyptian pyramids and obelisks—seemed to be the basis of the great memorials that have kept their significance and dignity across time."

In cross section, the arch would be an equilateral triangle, and it would be hollow, tapering from 57 feet at the base to 17 feet at the top. The arch would consist of 142 pie-shaped sections, each weighing approximately 50 tons. Each segment was to be tapered from bottom to top so that it would fit exactly on the segment below and be welded into place.

Fig. 37 The Gateway Arch. Courtesy of the Jefferson Expansion Memorial Park

Saarinen had created the "perfect architectural feat." It would be a beautiful structure of a shape never before built; it would be the tallest monument in the United States; it would be clad with an unprecedented amount of stainless steel; and it would last 1000 years and never need painting. In doing so, Saarinen had also designed a structure that men might not be able to build, especially because the structure would need to be built on a slant. (In 1907, when the Quebec cantilever bridge was being built across the St. Lawrence River, it fell during construction. When construction resumed in 1916, the bridge failed a second time due to faulty design.)

It should also be mentioned that Saarinen planned to have a 1079-step stairway inside the structure, so that visitors would be able to reach an enclosed observatory at the top of the arch. Although Saarinen's plan was submitted in 1947, construction did not actually begin until 1962.

In the meantime, Saarinen had second thoughts about the stairs and the likelihood that few visitors would attempt the climb to the top, which might take an hour. Otis Elevator was not interested in designing and building an elevator for the arch, but Saarinen finally found Richard Bowser, a one-time high school dropout, who had designed some elevators for parking garages. Within two weeks, Bowser submitted a design for trams that would ride up rails on the inside of the two legs of the arch. There were eight round capsules connected to each other on each of the two tramways. Each small capsule would seat 5 persons, for a total of 40 passengers per trip. Tall people had to keep their heads down, so cramped were the quarters. The capsules were gimbaled so that they would stay level. The ride time would be four minutes.

Groundbreaking for the arch took place in 1962. The excavation went down 60 feet before the pouring of the concrete began. The local Laclede Steel Company had the contract for steel plates and steel concrete reinforcing bars for the project. Al Weber, Laclede's chief metallurgist, became a self-appointed photographer of the construction. He took many trips up onto the arch during its construction and was soon on a first-name basis with many of the workers. Weber gave illustrated talks about the building of the arch and related an amusing story. "When the arch was about half way up it was decided to stretch a safety net between the two legs. After the net was in place a workman looked down and his hard hat fell right into the net but bounced right out again."

The Pittsburgh-Des Moines Steel Company at Neville Island, Pittsburgh, became the subcontractor for the construction and erection of the arch. The plan was to have 142 of the pie-shaped wedges fabricated at the Neville Island Plant. The 426 stainless steel plates would be supplied by U.S. Steel in Pittsburgh and Eastern Stainless Steel Company in Baltimore. The plates were type 304 stainless steel, ¼ inch thick, with a No. 3 finish. The largest wedges for the base of the arch would be 57 feet on a side and 12 feet tall. The top wedge would be 17 feet on a side and 12 feet tall. Each wedge would be tapered to fit exactly on top of the wedge below. The very large pieces were shipped by barge to St. Louis, the only way possible to have shipped such large pieces.

Construction required the unusual use of 50 ton creeper cranes that rode up the sides of the arch on rails as each segment was welded into place. Workmen were hoisted up in elevator cages.

When the arch was completed, except for the final eight-foot keystone segment, there was just a two-foot gap between the two legs. It was expected that the legs would have to be separated, and this was done with a powerful 250 ton hydraulic jack so that the final eight-foot keystone segment could be inserted.

On October 29, 1965, the arch was opened to visitors, but Saarinen did not live to see his masterpiece. He had died four years earlier of a brain tumor at the age of 50. Visitors wanting to reach the observation deck initially had to use the stairway. The trams were installed three years later, in 1968.

The cost of the Arch was $16.5 million, including $3.5 million for the transportation system. Approximately 1 million people take the trams to the top each year. In operation for over 30 years, the trams have traveled 250,000 miles, carrying 25 million passengers.

CHAPTER 10

History of Stainless Steel Melting and Refining

"Melting stainless steel is a highly-developed art; even at the present day artists are prone to guard their techniques quite jealously."
Carl A. Zapffe, 1949

THE TWO PRINCIPAL alloying elements in stainless steel are chromium in straight chromium steels and chromium and nickel in austenitic steels. Chromium is the difficult material to deal with, because it "gobbles up," as one writer put it, both carbon and oxygen. For these reasons, stainless steels are among the most difficult alloys to melt and refine. Nickel does not pick up either carbon or oxygen. A brief description of the three-step process generally used for the first 40 years is as follows:

1. *Melting:* Load arc furnace with low-phosphorus steel scrap, mill scale, or iron ore.
2. *Decarburizing:* Remove oxidizing slag and replace with a reducing or finishing slag of lime and fluorspar to which pulverized ferrosilicon is added from time to time.
3. *Alloying:* Add ferrochromium (high- or low-carbon ferrochromium) in two or three batches. Remove slag and pour.

The sources of chromium for smelting are chromium in the form of chromite ore or ferrochromium, an iron-chromium alloy that is available as a high-carbon type or a much more expensive low-carbon type.

It is also possible to use stainless scrap to replace all or part of the ferrochromium.

The Wild Process

Alwyn and Ronald Wild arrived in America from Sheffield, England, in 1926. They presumably had learned quite a lot about stainless steel, being from the center of the British stainless steel industry, and brought with them a new melting process. They bought the Hess Steel company, a defunct steel mill in Baltimore, Maryland, and called their firm the Rustless Iron Company, obviously intending to concentrate on making a low-carbon chromium alloy.

The Wild process followed the first two steps of the normal practice described previously. However, instead of adding ferrochromium in the third step, the oxidizing slag was replaced with chromium ore (chromite) and ferrosilicon. Metallic chromium was supposed to be reduced from the ore with silicon and alloyed with the melt without excessive silicon pickup.

After making dozens of trial heats, it seems that the carbon content was rarely low enough for rustless iron. The Wilds had no saleable product, and they went bankrupt in 1930. They were apparently never heard from again. Clarence Ewing Tuttle bought the Rustless Iron Company.

The Rustless Process

A.L. Feild (Fig. 15), working along lines similar to the Wild brothers, developed the essential principles of what would become known as the rustless process at Canton, Ohio, in 1926. It was a process for making low-carbon chromium stainless steel that promised to be much less expensive than the existing process for producing that material, which was to use low-carbon ferrochromium.

Feild applied these principles successfully in a pilot plant operation using a two ton electric arc furnace of the Heroult type in 1927. (Paul Hèroult, a Frenchman, invented the electric arc furnace for steelmaking in 1900.) A Heroult-type furnace is shown in Fig. 38.

In 1929, Feild produced rustless iron commercially in a six ton furnace at Lockport, New York. At that time, the term *rustless iron* referred to what is now called ferritic stainless steel. Later, Clarence Tuttle brought Feild to the Rustless Iron Company plant in Baltimore,

Fig. 38 Six ton Heroult-type furnace. Source: A.L. Feild, Manufacture of Stainless Iron from Ferrochromium, from Scrap, or from Ore, *Metal Progress,* Feb. 1933, p 15

where Feild proceeded to set up what would become a major stainless steel company.

From a metallurgist's standpoint, the rustless process is outstanding for three reasons:

- The unusual use of chromite (chromium ore) as a hearth liner
- The superheating to high temperatures to produce a major reduction of carbon with a minimum loss of chromium
- Returning chromium, which would otherwise be lost in the slag, to the melt by the addition of ferrosilicon

The name of the company was changed to The Rustless Iron & Steel Company. The company, after considerable litigation with the American Stainless Steel Company and the expenditure of half-a-million dollars, finally got a patent on their low-cost process. The patent was still in force in 1939.

An overview of the process is described by Feild in the article "Manufacture of Stainless Iron from Ferrochromium, from Scrap, or from Ore" in *Metals Progress*, February 1933, p 13. In a section titled "Future Expectancies," Feild concludes the following:

> "It is within the bounds of reasonable expectation that economic pressure and competition between the several melting processes will soon bring about some balance between metallurgical inventiveness and skill, on the one hand, and those fundamental items of basic cost which characterize all metallurgical melting processes. Even then the high chromium steels will always remain considerably more costly than ordinary steel. But if it is possible to produce steel ingots containing 0.10% carbon and 18% chromium, for instance, at a manufacturing cost of from 2½ to 4¢ per lb., there is no reason why the subsequent processes of heating and rolling to the finished sheet, strip, or bar should not be conducted on a scale sufficiently large and efficient to justify a selling price which would open up markets and applications not dreamed of except by believers in the philosophers' stone."

The imaginations of practical men were brimming with dreams.

The Linde Argon-Oxygen Decarburization (AOD) Process

In 1954, William A. Krivsky received a doctorate degree in metal refining from the Massachusetts Institute of Technology. He immediately was offered a job in the Metals Research Laboratory of the Union Carbide Corporation at Niagara Falls, New York.

Krivsky, in a settling-in period, apparently had no specific assignments and became interested in reading papers by researchers in England on the refining of stainless steel. He found that a recent thermodynamics study by Richardson on the relationship of chromium to carbon in stainless steel showed different results from an earlier study of the same subject by Hilty and Crafts. They had blown small induction heats of iron-chromium-carbon with oxygen in an attempt to establish equilibrium values for chromium and carbon.

Krivsky decided to resolve the difference between the two studies by extending the range of the chromium-carbon relationship studied while following the general method of Hilty and Crafts.

His initial work was in a 100 pound induction furnace. When blowing oxygen into the melt, he found that he could not control the high temperatures being produced by the oxygen because of the exothermic reaction. He conducted more tests where he diluted the oxygen with argon and immediately noticed that the carbon level was much lower than had been predicted by either Richardson or Hilty and Crafts. Krivsky knew he was on to something big.

Further lab work confirmed Krivsky's results, and he could see the possibility of a new steelmaking process that might drastically reduce carbon contents in stainless steel. There was a growing demand for the extra-low-carbon (ELC) stainless grades, but production costs were nearly prohibitive because the high finishing temperatures resulted in poor reproducibility and reduced the life of the refractory furnace lining. The argon-dilution idea offered a method for producing these extra-low-carbon grades, which was necessary to minimize damage to the steel by carbide precipitation.

Krivsky filed a patent application for the AOD system on June 27, 1956. The patent was contested and not issued until May 24, 1966, ten years later. The delay, however, proved to be fortunate, because it took 11 years to perfect the AOD process.

As an aside, it is interesting to note that the basic oxygen furnace steelmaking process for carbon and alloy steels was rapidly rising and leading to tremendous increases in oxygen production by the air liquefaction process. With minor changes, it was also possible to recover argon from air at a reasonable cost. Air contains almost 1% argon, which is an inert gas a little heavier than oxygen. These considerations led to the pursuit of what came to be called the Linde AOD process. Linde Air Products, a major supplier of oxygen and argon gases, was a subsidiary of Union Carbide.

Krivsky scaled up the process from a 100 pound to a 1 ton furnace, with generally favorable results. The work was done following the conventional practice of making stainless steel in an arc furnace, except for injecting the melt with argon-oxygen mixtures instead of pure oxygen.

Tests at Haynes Stellite. It was then decided to test the process in the three and five ton furnaces that were available at the Haynes Stellite Company in Kokomo, Indiana. They were then a division of Union Carbide. In 1911, Elwood Haynes had invented Stellite and stainless steel in Kokomo, about the same time that Harry Brearley had invented stainless steel in England. Haynes set up the Haynes Stellite Company but never became involved with the stainless steel business, except as a director of the American Stainless Steel Company.

Problems were encountered from the start in getting the argon and oxygen distributed uniformly throughout the melt using lances of various configurations to inject the gases. Although lancing with oxygen was effective when blown on the surface of the bath, it was found that the argon needed to be injected. When injection was tried, there was considerable splashing of the metal and erosion of the lances, despite attempts to cover them with refractory coatings of various sorts. It was thought that these problems could be solved in a furnace larger than the five ton furnaces at Haynes.

Union Carbide now needed a partner with larger furnaces and an interest in undertaking a joint venture to explore a new steelmaking process. Inquiries among all of the large stainless steel producers failed to arouse any interest whatsoever, but Joslyn Stainless Steel, a small producer in Ft. Wayne, Indiana, decided to take up the challenge. Union Carbide and Joslyn signed a contract in 1960.

Krivsky realized that his work with the AOD process was finished. He accepted an offer to become Vice President and General Manager of Brush Beryllium, a company in Reading, Pennsylvania. Krivsky was then approximately 30 years old.

Experiments on AOD Continue at Joslyn Stainless Steel. The campaign to develop an AOD process was in its sixth year. At Joslyn, the experiments were scaled up from the 5 ton furnace tests at Haynes Stellite to tests in a 15 ton furnace. However, the first tests conducted were utter failures. The oxygen and argon could not be dispersed evenly throughout the melt. A great deal of attention was paid to the lance problem, but no successful heats were ever produced because the surface area was far too large to obtain uniform distribution of the gases throughout the melt. Large-scale testing was discontinued, and work was continued on a laboratory scale.

It was decided that the only solution was to build a separate refining vessel. The first was a ¾ ton vessel that allowed for top blowing with oxygen and bottom blowing with argon through a tuyere. When refining was tried, the results were excellent and supported all of the laboratory work done earlier. The oxygen reduced the carbon content to a very low level.

However, there was great reluctance to the thought of requiring a second vessel to apply the argon-oxygen process. Going back to the arc furnace, they tried 45 more 15 ton heats using various lance configurations but to no avail.

Union Carbide and Joslyn finally decided to build a separate refining vessel. What was not realized at the time was that a second vessel would double the melting capacity. A 15 ton furnace with a simple multiple tuyere was built in 1962. The next five years was a period of endless testing and failure. There were many times when it was decided that it was useless to continue, but they kept on and on.

On October 24, 1967, the first full heat of stainless steel to successfully use the AOD process was produced. The refining time was only 58 minutes. The carbon content was an unbelievably low 0.008%, the chromium content was 18.63%, and the oxygen was 120 parts per million. It was an achievement that had taken 13 long years.

Joslyn Stainless Steel then built a full-scale commercial installation in 1968. The refining vessel was 9 feet in diameter and 13½ feet high. There were two tuyeres installed on the lower back side of the vessel. During 1968, 100 heats were made in the unit, with the following conclusions reached:

- "The process economies were established beyond any question."
- "The charge of the electric furnace could be made of the lowest-cost chromium and nickel without any restrictions on the starting levels of carbon or silicon."
- "Recovery of chromium was very high, around 97%."
- "Silicon usage to recover oxidized metals was reduced by 40%."

Joslyn then modified this first production unit and converted their entire production to the AOD process in July 1969. The following December, Joslyn metallurgists reported the results of 1300 heats at the annual Electric Furnace Conference in Pittsburgh.

Decarburization by the AOD process meant that ELC alloys (with 0.03% carbon or less) could be produced, thereby reducing available carbon for any significant precipitation of chromium carbides during welding, heat treatment, or service at temperatures between 900 and 1500 °F. Less precipitation of chromium carbides meant that enough chromium remained in solid solution and thus was available to form chromium oxides for corrosion protection of the steel. In particular, the ELC alloys reduced carbon levels enough that free carbon in the grain boundaries of steel did not deplete solid-solution chromium in the grain-boundary region. Other advantages of the AOD process were:

- The productivity was doubled.
- Lower refining temperatures reduced refractory costs.
- Elimination of the refining and finishing operations from the arc furnace more than offset the operating costs of the AOD.
- The yield of the metallic alloying elements is increased.
- Lead is kept at very low levels of no more than 0.007%.
- Sulfur, oxygen, hydrogen, and nitrogen levels are far lower than previously possible.
- The consistency, control, and reproducibility improved ductility, fatigue strength, and the machinability of many alloys.
- It is possible to add nitrogen within very close limits for producing new nitrogen-hardening alloys.
- The refinements of the process open the possibility of developing new alloys.

Starting the AOD Business. Early in 1970, Union Carbide and Linde contacted virtually every major stainless steel producer in the world, offering licenses for the new process. Demonstrations were also offered at Joslyn's Ft. Wayne plant. Despite the advertised advantages of the AOD process, it took a surprisingly long time to find customers. The addition of another step to the production process and the idea of paying a fee for every ton of steel produced were obstacles not easily overcome.

Union Carbide had invested millions and spent 13 long years developing what would eventually become one of the most significant steelmaking discoveries of the 20th century. Joslyn, of course, got their license right away, in April 1968. They were eager to show what they had accomplished.

In January 1970, Haynes Stellite became the second licensee, followed by Illsa Viola in Italy in July 1970, Electraloy of Oil City, Pennsylvania, in September, and Eastern Stainless Steel in Baltimore, Maryland, in December 1970. Only five licensees through 1970 was a surprisingly low number, but in 1971 there was an additional eleven—six American and five foreign.

In 1973, Union Carbide predicted that 50% of the stainless steel would eventually be made by the AOD process. They were wrong. It turned out to be closer to ⅔. By 1982, 79 AOD vessels had been installed in 60 steelworks throughout the world, and 24 foundries had installed an additional 24 vessels. By the year 2000, over 100 AOD vessels had been installed.

Other Steel-Refining Developments. The development of the AOD process led to the development of other types of converters. These were important developments, but none became as successful as AOD. The processes included:

- KCB-S process by Krupp
- K-BOP process by Kawasaki Steel
- K-OBM-S process of Voest Alpine
- MRP process of Manesmann Demag
- CLU process of Creusot-Loire-Uddeholm
- Sumitomo STB process
- TMBI process of Allegheny Ludlum
- VODC process of Thyssen
- AOD/VCR process of Daido

There were also three-stage triplex processes using the electric arc furnace, a converter for preblowing, and a vacuum decarburization unit for final refining.

CHAPTER 11

Two New Classes of Stainless Steel

UNTIL 1930, the three classes of stainless steel were the martensitic, ferritic, and austenitic (Fig. 39–41). The austenitic grades, such as 18-8, became the most popular for several reasons: their ease of production, fabrication, and welding and their superior resistance to corrosion in most environments. On the other hand, the austenitic alloys were considerably more expensive than the martensitic and ferritic grades and were susceptible to intergranular corrosion and stress-corrosion cracking.

Duplex Stainless Steel

In 1927, Bain and Griffiths at U.S. Steel prepared phase diagrams of the iron-chromium-nickel system and described austenite-ferrite alloys with 23 to 30% chromium and 1.2 to 9.7% nickel, but no information on properties was presented in their paper. However, their paper apparently came to the attention of metallurgists at the Avesta Ironworks in Sweden, who developed two ferritic-austenitic alloys with the main objective of reducing the problem of intergranular corrosion that was prevalent with the austenitic alloys. They produced Avesta grade 453E, which was a chromium-nickel alloy, and grade 453S, which was the same as grade 453E but also contained 1.5% molybdenum. These alloys had approximately the same carbon content (0.1%) as the austenitic grades but showed definitely improved resistance to intergranular corrosion. They also had equal or better resistance to uniform corrosion than the austenitic grades.

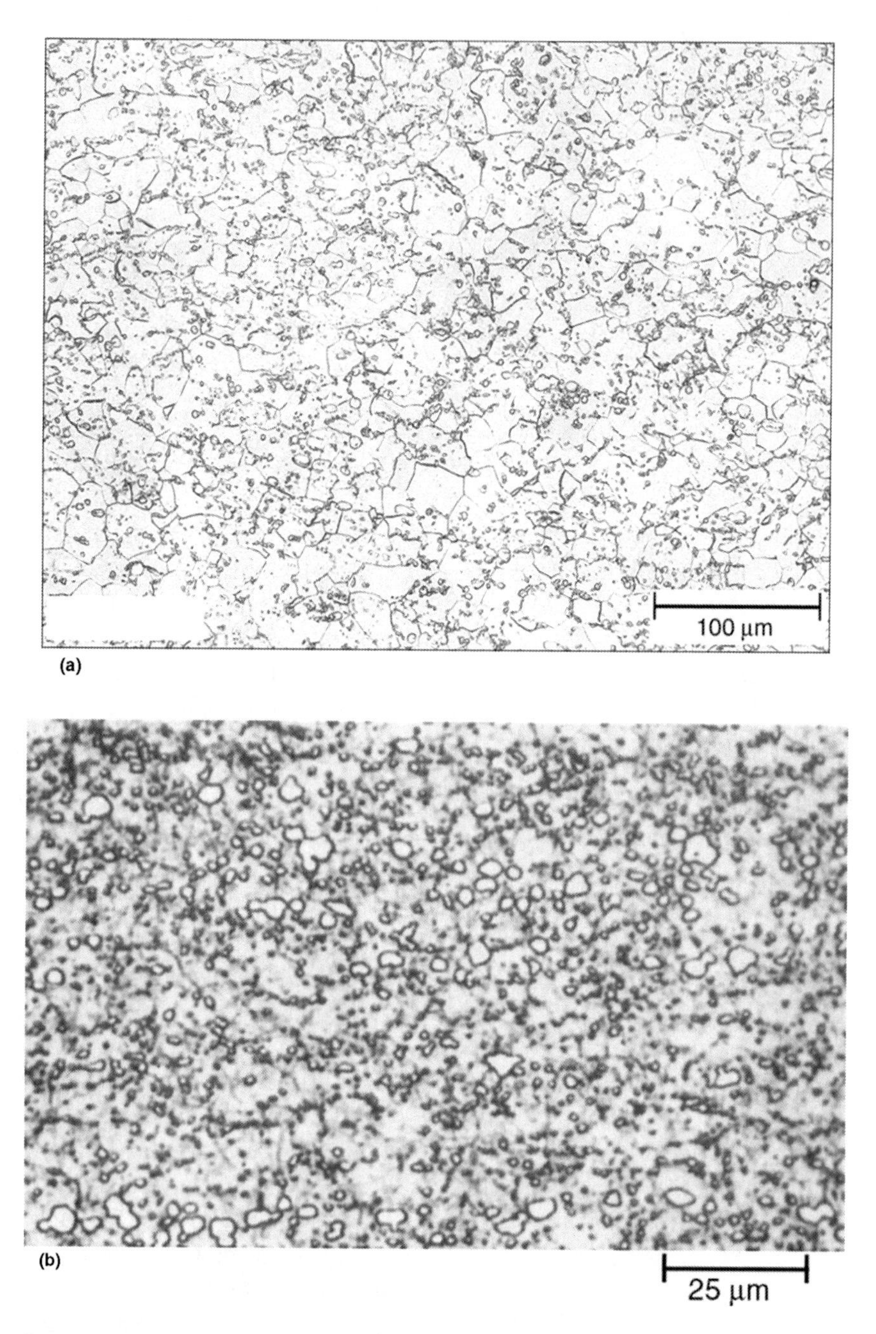

Fig. 39 Microstructure of a martensitic stainless steel (type 410; UNS number S41000). (a) Annealed. (b) Tempered after hardening

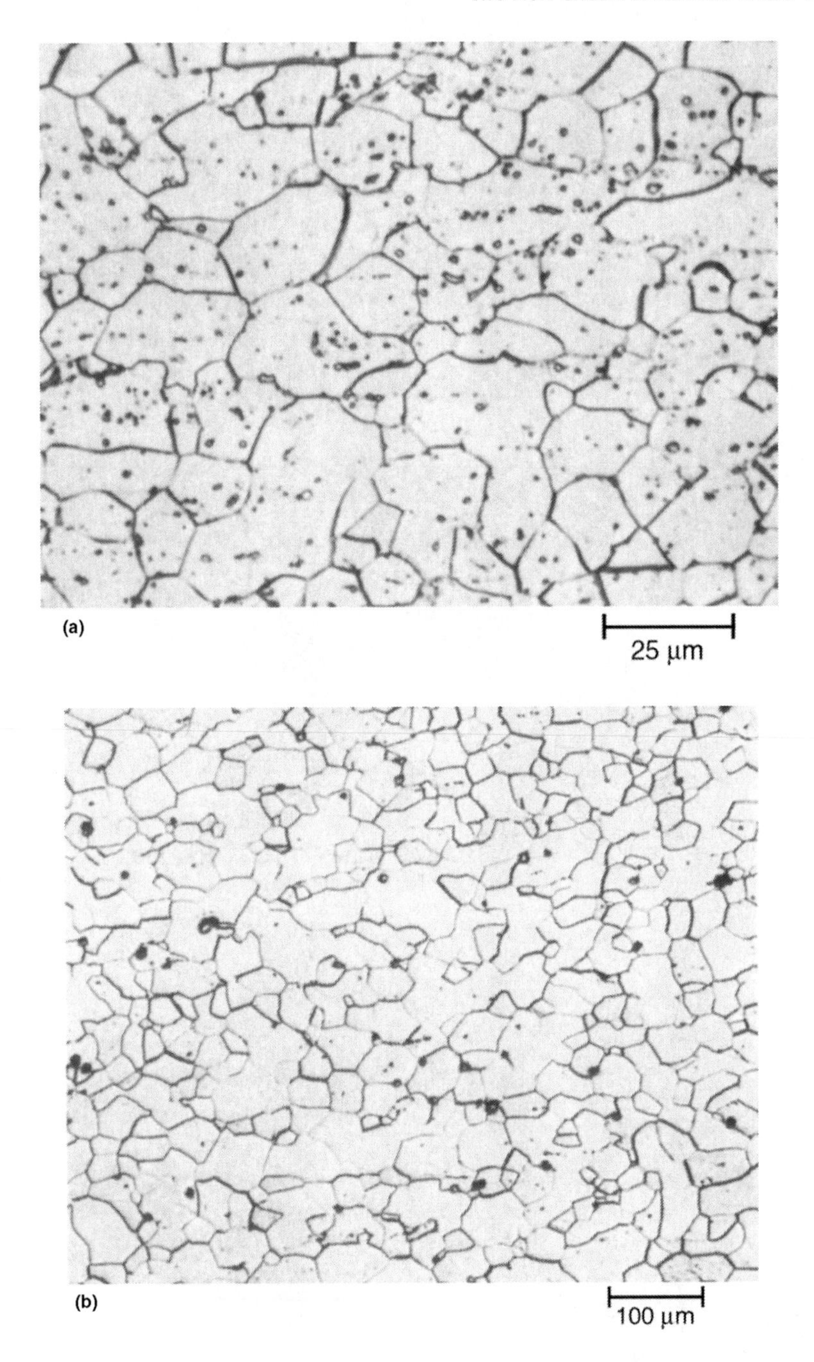

Fig. 40 Microstructure of two ferritic stainless steels. (a) Type 409 (UNS number S40900) muffler-grade strip in annealed condition. (b) Type 430 (UNS number S43000) annealed strip

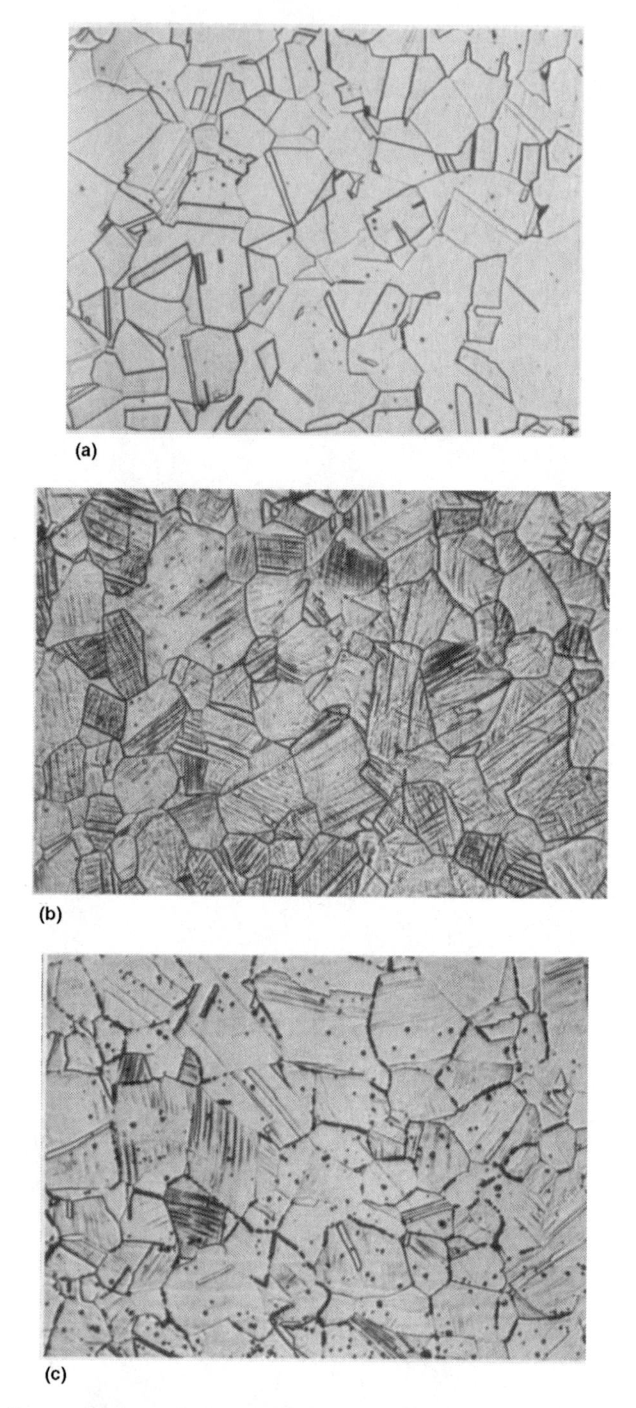

Fig. 41 Microstructure of 304 austenitic stainless steel from three specimens of a fabricated part from welded strip. (a) Annealed location that was unaffected by processing. (b) Region with slip bands caused by roll forming. (c) Heat-affected weld zone with carbides in the grain boundaries. Source: *Atlas of Microstructures of Industrial Alloys,* Volume 7, *Metals Handbook,* 8th ed., American Society for Metals, 1972, p 135

The positive effect of the duplex microstructure (Fig. 42) with regard to intergranular corrosion was reported by Payson and Harrison in 1932 and by Lindh in 1934.

The first ferritic-austenitic stainless steel contained 60 to 70% ferrite after quench-annealing at 1000 to 1050 °C. The duplex microstructure more than doubled the strength of the annealed austenitic grade 304, as shown in the following table:

Grade	Yield strength, MPa	Tensile strength, MPa	Elongation, %
453E	510	593	35
453S	640	715	21
304 (annealed)	207	517	40

The Avesta 453S grade, which was later adopted as AISI grade 329 and Swedish SIS 2324, had a yield strength three times that of type 304 (annealed) and a tensile strength 1.4 times that of type 304. It has only been since the 1970s that duplex stainless steels have become popular, primarily by the increased use of the argon-oxygen decarburization refining process and the continuous casting of slabs.

Duplex stainless steels have been classified according to the first period (1930–1960) and second period (1960–1990). The designations for these duplex alloys in the United States have been primarily according to Unified Numbering System (UNS) numbers in the S3*xxxx* series. The only grade that was given an American Iron and Steel Institute (AISI) number was AISI 329, which is now also UNS S32900. As of 2008, 17 wrought duplex grades of stainless steels had been adopted by ASTM International. The cast equivalent of type 329 is CD-4MCu in the Alloy Casting Institute system and J93370 in the UNS system.

Duplex stainless steels have some of the highest strength levels of any of the corrosion-resistant materials. The mechanical properties are improved by balancing the structure to be approximately half ferrite and half austenite, with the ferrite contributing the improved mechanical properties. The greatest amount of development work in recent years has been with these new alloys.

Precipitation-Hardening Steel

The fifth and last of the major stainless steel classifications is the precipitation-hardening or age-hardening group, also known as PH

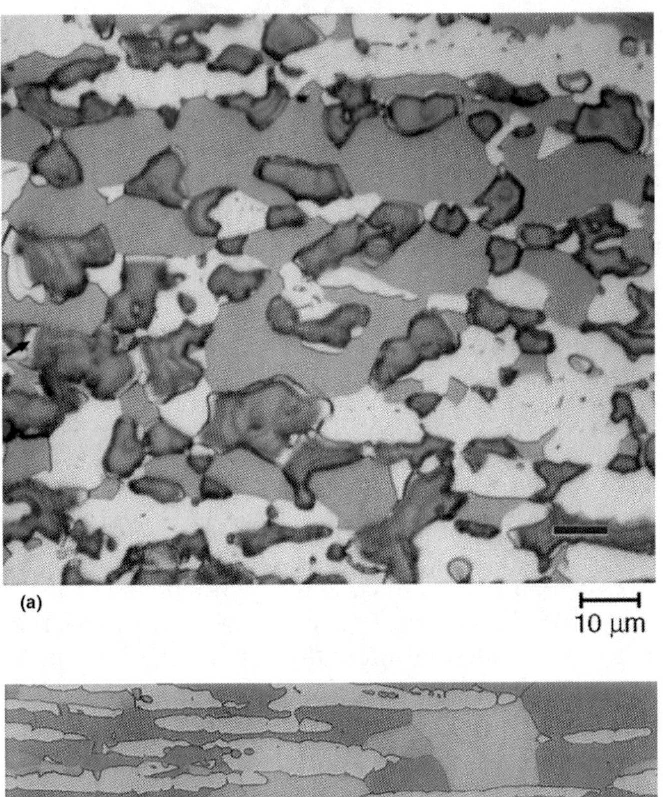

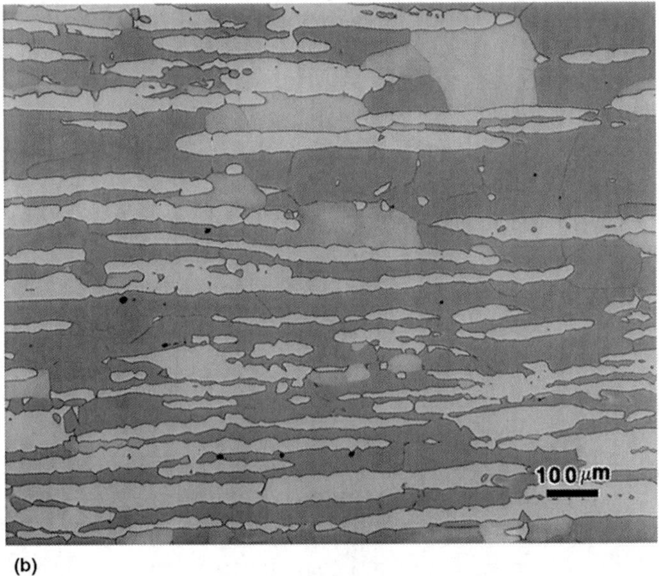

Fig. 42 Microstructure of two duplex stainless steels. (a) Solution treated and aged 7-Mo (Fe; <0.1 C; 27.5 Cr; 1.5 Ni; 1.5 Mo). Ferrite is light gray; sigma phase is dark; austenite is the whitest shade. (b) Type 2205 duplex wrought stainless steel (UNS number S31803). Ferrite (darker) and austenite (white). Composition is balanced to produce approximately equal amounts of ferrite and austenite at room temperature. Courtesy of Buehler Ltd.

steels. Unlike the alloys in the other classifications, these steels are not characterized by their crystalline structure but rather by a particular strengthening mechanism in one of the structures.

A steel of this type was first discovered by William Kroll of Luxembourg in 1929. His alloy was strengthened by small additions of titanium, which combine with carbon to produce fine titanium carbide precipitates that strengthen the crystal matrix. Kroll later became famous in the 1940s when he developed the first commercial process for refining titanium and zirconium.

The first precipitation-hardening alloy to be produced commercially was Stainless W, an alloy developed by the Carnegie-Illinois Steel Company of Pittsburgh. This was followed in 1948 by Armco's development of 17-4 PH and later by 17-7 PH and PH 15-7 Mo. Allegheny Ludlum Steel Company developed A286, AM350, and AM355. The Carpenter Steel Company developed Custom 450, Custom 455, and Custom 630. These alloys contain 11 to 18% chromium, 3 to 27% nickel, and smaller amounts of additional metals, including aluminum, copper, columbium (niobium), molybdenum, titanium, and tungsten.

The alloys are characterized by having extremely high strength and corrosion resistance as good as or better than type 304 stainless steel. They achieve their high strength by the formation of minute precipitates during an aging treatment. The alloys are finish-machined in the solution-treated condition, followed by hardening, typically by a four hour aging treatment at 1000 °F (540 °C).

Because the names of these proprietary alloys could not be used in ASTM specifications, several numbering systems were devised for this purpose. It was the practice of AISI not to assign numbers to proprietary alloys, but a group, now unknown, decided to start numbering these alloys and others with a three-digit 600 series of numbers. Because the AISI used a three-digit system, it was assumed by many that the 600-series numbers were AISI numbers.

Numbers 630 to 635 were assigned to some of the PH steels, and others were assigned XM numbers, XM-9, 12, 13, 16, and 25. The designations are all used in ASTM specifications rather than the trade names. The alloys are also identified by UNS designations.

Some of the PH alloys are martensitic, some are semiaustenitic, and alloy A286 is austenitic (Fig. 43). The alloys are used primarily for forgings and fasteners for aerospace and military applications.

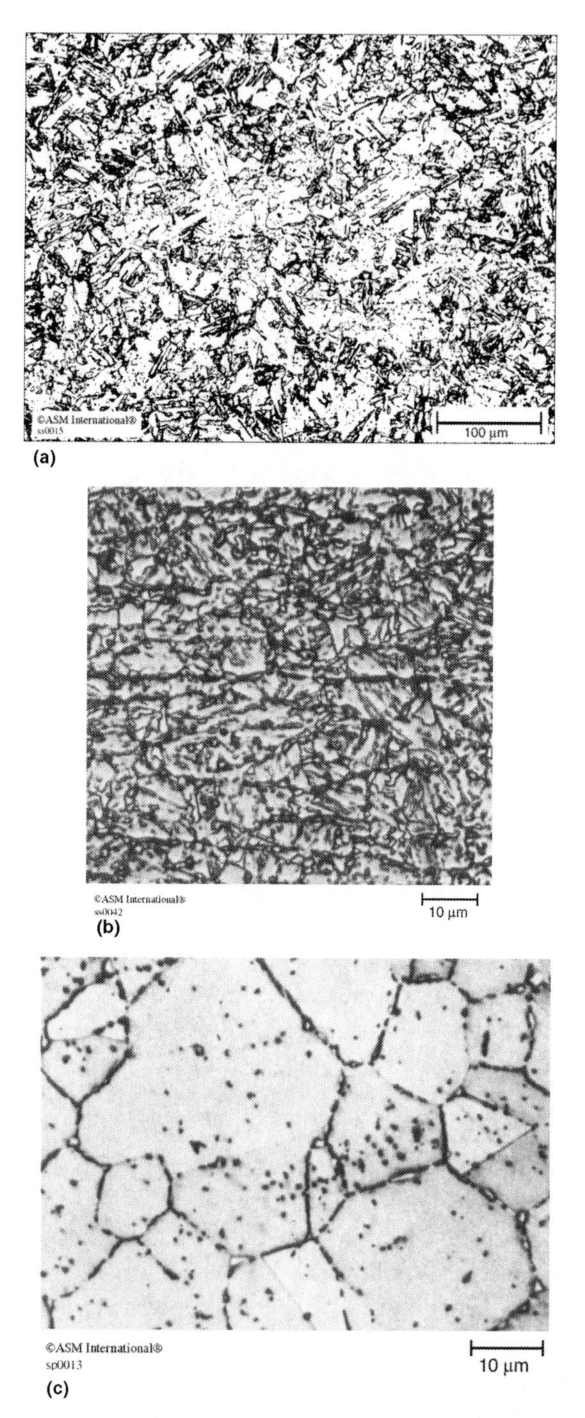

Fig. 43 Microstructure of precipitation-hardening (PH) stainless steels. (a) Martensitic PH stainless steel type 15-5 PH (UNS number S15500) in solution-treated and aged condition. (b) Semiaustenitic PH stainless steel type 17-7 PH (UNS number S17700) in solution-treated and aged condition. (c) Austenitic PH stainless steel type A286 (UNS number S66286) in annealed condition

CHAPTER 12

Stainless Steel Applications

STAINLESS STEELS have a wide variety of applications for household products, food-handling equipment, major appliances, medical equipment, and industrial equipment. Stainless also is featured in many architectural designs and monuments. Many of the most important applications of stainless steel can be found in the transportation industry, where both the cutlery martensitic and the chromium-nickel austenitic stainless steels have been used.

Household Products

Cutlery and Tableware. When Harry Brearley discovered his iron-chromium alloy in the city of Sheffield in 1913, one of his first ideas was that the metal may be useful in the cutlery trade. Sheffield was the center of the cutlery trade in England and had been so since the 14th century. Brearley was well aware that the cleaning and sharpening of knives was a continual household chore, principally because the steel knife blades became stained with rust that had to be cleaned with emery.

Brearley reasoned that his stainless steel could be hardened and sharpened to a fine edge, just like carbon steel, but it wouldn't rust. In time, Brearley's alloy was used to make all the knife blades in Sheffield. His steel was known either as cutlery steel or stainless steel. He arranged to have each blade stamped with the words "Firth-Brearley Stainless Steel," a practice that lasted for ten years.

Although Brearley's steel was used for carving knives, sterling silver and silver plate were preferred for forks, spoons, and knives for table settings, because they could be crafted with ornate patterns, which became the vogue during the first part of the 20th century in America.

In Europe, however, tableware was being made of the chromium-nickel stainless steel, also known as 18-8, in Solingen, the cutlery center of Germany, probably as early as the 1920s. Solingen stainless steel tableware was marketed in the United States beginning in approximately 1945.

In the mid-1950s, the big American silversmiths began to close down their silver-plated tableware lines to get into the stainless steel tableware business. The patterns were rather simple at first, because the stainless steel was much more difficult to work.

In the early1960s, the Oneida Silver Company, Oneida, New York, made a definite commitment to stainless steel flatware. It was a bold and pioneering move, considering the fact that stainless steel flatware had a secondary position in the industry, lagging far behind the more expensive and more prestigious sterling and silver-plated flatware. In the mid-1960s, Oneida dramatically changed the quality and appearance when they introduced "Chateau," one of the first stainless patterns to emulate sterling silver designs. By the late 1960s, Oneida's technological advances enabled them to introduce the first ornate traditional pierced pattern in stainless steel (Fig. 44). As stainless steel flatware gained a place in the silver departments of more and more fine department stores, its attractive appearance and ease of care made it more popular than silver plate or sterling. This rise in the popularity of stainless sparked Oneida's recovery from what had been a severe slump in their business. By 1983, Oneida was selling over half of all flatware purchased in the United States.

Kitchenware. Polarware, the first stainless steel cookware in America, was manufactured by The Polar Ware Company, Sheboygan, Wisconsin. It was primarily for commercial use. The first known public display of stainless steel kitchenware in America was at the Chicago Centennial Exposition of Progress in 1933 to 1934 (Fig. 45). The exhibit was sponsored by the United States Steel Corporation and included items that were made of the new 18-8 stainless steel alloy. All articles were gleaming and included pots and pans, a bowl, colander, tea kettle, water pitcher, double boiler, baking pan, ladle, spatula, counter tops, and a sink.

The deep-drawn items, such as the pan, bowl, and pitcher, demonstrated the excellent formability of the material, something that could not have been achieved with the cutlery type of stainless steel that had arrived in America almost 20 years earlier.

On close inspection, it was noticed that there was no display of a particularly common kitchen utensil, the frying pan, and for good

Fig. 44 Stainless steel Oneida tableware circa late 1960s. Courtesy of D. Gymburch

reason. The thermal conductivity of the 18-8 variety of stainless steel is relatively poor, so that the heat from a gas flame would not spread evenly. Cladding other more conductive metals with stainless steel would eventually deal with that problem, with varying degrees of success.

The 18-8 austenitic stainless steel was an excellent material to be used for cookware. It was beautiful and durable. Its hard, smooth surface was readily cleaned. The taste and appearance of food cooked in these utensils was unchanged, and the metal was not affected by contact with acids from fruit and vegetables.

The first stainless steel cookware that was produced with superior heat distribution was made in the 1930s by the Stainless Metals Com-

Fig. 45 U.S. Steel display of stainless steel kitchenware at the Century of Progress exhibition during the 1933–1934 Chicago Centennial Exposition

pany of New York City. The stainless steel cookware had a carbon steel core that conducted heat more rapidly than stainless steel.

The Revere Copper & Brass Company, Rome, New York, introduced Revere Ware at the Chicago Housewares Show in 1939. It was the cookware with the copper-plated bottoms for heat distribution. There was a complete line of many types of cookware, as well as canisters of many sizes. The Revere Brand became the most profitable division of the Revere Copper & Brass Company.

The first stainless-clad copper core cookware was made by Lalance and Grosjean of New York and the Stainless Steel Company of America, Walled Lake, Michigan. The copper core provided improved heat distribution.

In the 1950s, the Housewares Division of G.S.W., Ltd., Baie Urie, Quebec, brought out a line of three-ply cookware with 18-8 stainless steel on both sides of a carbon steel core. The idea was that the steel core would help distribute the heat because of its better thermal conductivity. The tri-ply material was introduced by Allegheny Ludlum Steel Company especially for the cookware market.

All-Clad Metalcrafters. In 1960, John Ulam established Composite Metal Products at Canonsburgh, Pennsylvania, for the purpose of producing clad metals for various companies. In 1967, Ulam formed an alliance with Alcoa to create Clad-Metals, Inc. Ulam was a metallurgist and was awarded 50 U.S. patents pertaining to bonded metals. He was also instrumental in the U.S. Mint's conversion from silver coins to the bonded layered metals in use today.

In 1971, Ulam began producing a high grade of finished stainless-clad aluminum cookware as All-Clad Metalcrafters at Canonsburg, Pennsylvania. In 1988, Sam Michaels, a Pittsburgh entrepreneur, purchased All-Clad and expanded the production of cookware.

In 1999, Waterford Wedgwood, the famous china and porcelain company headquartered in Ireland, purchased All-Clad and made substantial investments in the company.

In 2004, Groupe SEB purchased All-Clad to expand All-Clad product offerings to a global market. SEB is a Swedish company with 20,000 employees in 20 countries.

Food Handling

Stainless steels have become the premier materials for handling, processing, storing, and transporting foods. They have the required corrosion resistance to vegetables, fruits, juices, meat, fish, jellies and marmalade, dairy products, beer and wine, confectionary products, syrup, and molasses. They do not react with foods to cause them to change color or taste or to become contaminated in any way. Foods may be cooked, processed, stored, and transported in stainless steel equipment, and the metal may be easily cleaned and polished without reacting with cleaning products. It has completely replaced the use of other metals and coated metals, such as tinned copper.

The original cutlery grade of stainless that was introduced in America in 1915 was used fairly successfully at first in most of these applications. In 1923, the Firth-Sterling Steel Company of McKeesport, Pennsylvania, reported that "this steel is satisfactory for use with fruits, vegetables, meat and fish and the remarkable thing is that vinegar which contains five percent acetic acid has little or no effect upon properly hardened and polished stainless steel." Firth-Sterling specifically noted that potential applications were breadmaking machinery, cutlery, soda fountain parts, and dairy equipment.

The cutlery steels, however, were hard metals and not well adapted to fabrication, unlike the 18-8 chromium-nickel steel that American mills started to produce under German licenses beginning in 1927. The 18-8 austenitic steel had superior corrosion resistance, superior formability, and a handsome appearance when polished. However, the alloy had twice the alloy content and was about twice the cost of the cutlery type of stainless steel, approximately 50 cents per pound. People in the food-handling business would have to be convinced of the reasons to buy the more expensive material.

Companies such as the United States Steel Company were interested in getting into what appeared to be a large market for stainless steel equipment, but there was no book on the properties of the material as they pertained to food handling. U.S. Steel set out to remedy this by initiating an extensive testing program, turning one of their laboratories into a large kitchen.

U.S. Steel decided to look into the effects of salt concentration, acid level, dissolved oxygen, and stirring when cooking a wide variety of foods at different times and temperatures. The test program included the cooking of sauerkraut, tomatoes, spinach, gooseberries, coffee, tea, rhubarb, chicken, rice, cauliflower, potatoes, and fish.

After cooking, the containers were stored with their contents in refrigerators for 72 hours, after which the contents were removed to glass jars and tested for taste, color, and odor. The stainless steel containers were carefully weighed before and after the tests and were inspected for signs of discoloration, corrosion, or pitting. The food samples were then sent to the analytical laboratory of Titus, Elkins, Finn, Fairhall, and Drinker to determine the amounts of iron, chromium, and nickel in the samples. The amounts proved to be minuscule, and it was concluded that the presence of these small amounts could be disregarded. No changes in appearance, taste, or odor were detected. It was concluded that the 18-8 steel was entirely satisfactory for cooking foods of all types.

Use in Dairy Equipment. Until the first installation of stainless steel equipment, the selection of proper material for the dairy industry was a real problem. Dairy products are considered to be among the most sensitive and most easily contaminated foods. Certain metals, especially copper, copper alloys, zinc, and iron, are particularly objectionable in this regard.

Stainless steels were recognized and used for dairy equipment soon after the introduction of the formable 18-8 alloy. In the 1930s, 18-8 stainless steel was used in pasteurizing and cooling vats, milk handling equipment, and transportation. Trucks with 18-8-lined tanks transported dairy products from the farm to the dairies in the 1930s (Fig. 46).

Fig. 46 Milk truck with a 2700 gallon tank lined with 18-8 stainless steel. Source: *Food Handling Advances,* U.S. Steel, 1935

The 18-8 alloys have performed well in dairy use for many years. The material is fine for processing, including pasteurizing, storage, and transporting.

Other Food-Handling Applications. Stainless steels have been the material of choice for virtually all food-handling applications, including meat handling, the bakery industry, confections, and the brewery industry.

Architecture

Metal architecture is a relatively new phenomenon. Aside from roofing, the earliest use of metal in architecture was probably the 100 foot long Iron Bridge built over the River Severn at Coalbrookdale, England, in 1782. It was an arched bridge constructed by bolting flat cast iron bars together. The bridge is still standing. Other bridges were built of cast iron and wrought iron in the 19th century, as well as one steel bridge, the Eads Bridge over the Mississippi River, in 1874. Copper is another metal that has been used in architecture for many years, especially for roofing.

Although stainless steel had been discovered in 1912, it was not until 17 years later that anyone decided to use it in architecture. The first exterior use of stainless in architecture happened in 1929 when the famed London hotel, the Savoy, used a small amount of the metal to enhance the appearance of its entrance. The front of the canopy over the entrance was covered with a stainless steel sheet to which a large neon "Savoy" sign was attached. Stainless steel was also used on the façade around the entrance, making it quite handsome (Fig. 47).

The next and most remarkable use of stainless steel in architecture was that of the Chrysler Building, which is still generally regarded as the most outstanding building on the New York City skyline. On the occasion of the Symposium on Stainless Steel in Architecture, held in Berlin on June 15, 2005, Finnish architect Esko Meitten, in his presentation, cited the Chrysler Building as being "one of the icons of architecture."

At the time of construction, the 77-story Chrysler Building was to be the world's tallest building and was certainly one of the most highly decorated. The design, innovations, and the spire emerging from within the walls of the building to reach a record height of 1047 feet were truly dramatic accomplishments. In Chapter 7, "The Chrysler Building (1930)," the author referred to the building as "the miracle on 42nd Street."

Fig. 47 The canopy of the Hotel Savoy. Courtesy of the Nickel Development Institute (1999, Catherine Houska)

The most significant use of stainless steel in the Chrysler Building was for the 175 foot tower that tapered to a 10 foot tall spire at the top. The tower consisted of a series of seven arches (Fig. 48) that diminished in size from bottom to top. Within each arch were tall triangular windows on each of the four sides of the building. The lowest

Fig. 48 Arcs of the Chrysler Building. Courtesy of the Nickel Development Institute (Tim Pelling)

arch had seven windows, the next had six, and so on until there was only one window within the top arch. The surfaces surrounding each window were clad with Nirosta (18-8) stainless steel, so that each arch consisted of glass and metal.

Architect Van Alen described his building as follows: "With stainless steel the structural lines and metal facings are intensified by the mirrored surfaces reflecting the ever-changing light from the sky." Indeed, his detailed account (excerpted in Chapter 7) describes many functional and design advantages of stainless steel in architecture. Stainless steel was also used on the façade up through the third floor, the entrance (Fig. 49), exterior ornaments, and extensively for decoration in the lobby.

Fig. 49 Entrance to the Chrysler Building. Courtesy of the Nickel Development Institute (1996, Catherine Houska)

Van Alen's use of this attractive metal caught the attention of architects, but only a few other buildings were built using stainless steel during the Great Depression of the 1930s. These included the Empire State Building, completed a year after the Chrysler Building and setting a new record height of 1252 feet (382 meters) tall with 85 stories of office space. It was the world's tallest building for 41 years. The 200 foot (61 meter) dirigible mooring mast and five stories of office space were late design additions to surpass the Chrysler Building's height. The exterior is gray limestone ornamented with vertical strips of stainless steel (Fig. 50). The stainless steel spandrel panels start on the sixth floor and extend upward outside each window group to form a sunburst at the top of each tier. The 85th-floor observation level had stainless steel railing installed in 1949. The lobby of the building is also extensively decorated with stainless steel.

Another early use of stainless steel for exterior and interior applications was the 44-floor Philadelphia Savings Fund Society Building in 1932. A considerable amount of stainless steel was used in the building, at the entrance, and on the interiors and exteriors of the windows

(a)

(b)

Fig. 50 Stainless steel spandrel panels on the Empire State Building. (a) Undated photo from the 1930s. (b) 1996 (Catherine Houska)

of the Sun Room on the 53rd floor. The Executive Dining Room, on the same floor, had a kitchen with type 302 stainless steel sinks, work surfaces, and steam table. Type 302 was also used on the interior for railings, column covers, and the escalators that led to the actual bank on the second floor.

Although the building industry was stagnant during the Great Depression, the wonderful benefits of stainless steels were recognized for a host of interior and exterior applications. After World War II, more buildings were using stainless steel. The first significant stainless steel "curtain wall" application was in 1948 for the four-story office of the General Electric Turbine Plant Building No. 273 in Schenectady, New York. The curtain wall was corrugated, 0.038 inch (0.97 millimeter) thick type 302 stainless steel panels that were spot welded to the steel structural framing.

In 1954, an entire skyscraper, the Socony-Mobil Building (Fig. 51) on 42nd Street in New York City, was the first to have its exterior wall completely clad with stainless steel. This 42-story office building was the world's largest metal-clad building in 1954. Except for the windows, the entire building was sheathed with 7000 type 302 stainless steel panels. A considerable amount of stainless steel was also used in the lobby and for the elevator doors.

The panels for the exterior curtain wall were just a little over 1⁄32 of an inch thick and were press formed into a trihedral pattern to prevent waviness and to break up reflections. The appearance of the Socony-Mobil Building led critics to call it "The Waffle Building." However, in 2003, the buttelike building near Grand Central Terminal was named "one of New York's most striking skyscrapers" by the Landmarks Preservation Committee. The building was cleaned for the first time in 1995 (Fig. 52).

The cladding of buildings with stainless steel has since become common. In 1999, the largest known application of stainless steel as a curtain wall was for the Petronas Twin Towers Buildings in Kuala Lumpur. The buildings were covered with 700,000 square feet of stainless steel. The skyscrapers were the world's tallest at the time.

Hundreds of buildings and exterior structures around the world use stainless steel for both aesthetics and practical benefit. Photographs (see insert) show some noteworthy international stainless steel structures, including the dramatic Cloud Gate sculpture in Chicago's Millennium Park and the massive statue of Genghis Khan, located near Ulaanbaatar, Mongolia. Examples of stainless steel for exterior architectural applications include:

Fig. 51 150 East 42nd Street (formerly the Socony-Mobil Building) in New York City. Courtesy of the Nickel Development Institute (1996, Catherine Houska)

Pittsburgh Civic Arena	Pittsburgh, Pennsylvania	1961
Elephant & Castle Substation	Newington Causeway, London	1962
Kearns Communications Group Building	Dayton, Ohio	1973
Enfield Civic Centre	Enfield Borough, Middlesex, United Kingdom	1973
Pier Pavilion	Herne Bay, Kent, United Kingdom	1976
ICI Building	North York, Ontario, Canada	1981
Sun Life Centre	Toronto, Ontario, Canada	1984
Michael Fowler Centre	Wellington, New Zealand	1985

Fig. 52 First cleaning of the Socony-Mobil Building in 1995. Courtesy of J&L Specialty Steel

These buildings and structures (see insert) represent a range of applications and service environments. Some have had regular maintenance, while others have had none, as described in the brochure "Timeless Stainless Architecture" (C. Houska, P.G. Stone, and D.J. Cochrane, Reference Books Series No. 11 023, Nickel Development Institute, Toronto, 2001). This brochure describes 16 noteworthy, international, exterior stainless steel architectural applications that were completed between 1929 and 1985.

In addition to cladding of buildings, stainless steel has been an inspiration for elements of interior design, architectural façades, craftsmanship, and art (see insert). This appreciation of stainless steel was clearly recognized in the 1920s, and one of its admirers was Oscar Bruno Bach (1884–1957). Oscar Bach was an exceptional artist and craftsman who designed and executed many outstanding original works in metals crafts that are found throughout the world in museums, churches, temples, public buildings, and private residences. One outstanding example is an entrance door (see insert) with an expanse from pressed stainless steel. In 1938, a collection of Bach's stainless steel metalwork, including a door, its trim, and grillwork, was placed on permanent exhibit in the Procurement Division of the Treasury Department.

Fig. A Latticelike exterior on the IBM Buildings in Pittsburgh, Pennsylvania (Five Gateway Center). The loaded bearing trusses are sheathed with sheets of stainless steel. Architects: Cutis and Davis, New Orleans. Source: *New Horizons in Architecture with Stainless Steel,* American Iron and Steel Institute, 1965, with permission

Fig. B Water tower at General Motors Technical Center, Warren, Michigan. Constructed from stainless-clad structural steel plate ($^{1}/_{16}$ inch stainless steel on $^{1}/_{16}$ inch structural steel plate). Associated architects: Eero Saarinen; Smith, Hinchman, and Grylls. Photo: Baltazar Korab. Source: *New Horizons in Architecture with Stainless Steel,* American Iron and Steel Institute, 1965, with permission

Fig. C Water intake gate structures at Niagara Power Project, Niagara Falls, New York. The 100 foot high structure was sheathed with stainless steel. Source: *New Horizons in Architecture with Stainless Steel,* American Iron and Steel Institute, 1965, with permission

Fig. D Pittsburgh Civic Arena, Pittsburgh, Pennsylvania. When constructed, it had the world's largest dome and retractable roof. There are no interior supports. The stainless steel dome is 415 feet (126 meters) in diameter and consists of 7800 pieces of stainless steel that were joined with flat lock-and-batten seams to form eight movable leaves of the dome roof, which can be opened in two minutes. Courtesy of the Pittsburgh Civic Arena

Fig. E Elephant & Castle Substation, Newington Causeway, London. Curtain wall of type 316 stainless steel exterior wall panels with a pressed pattern. Courtesy of the Nickel Development Institute (1999, Catherine Houska)

Fig. F Kearns Communications Group Building, Dayton, Ohio, in 1999. Two sides of the building have type 304 stainless steel curtain walls with no windows. The other two sides are glass curtain walls and doors with solid stainless steel framing. Courtesy of Edward Madden, Kearns Communications Group

Fig. G Enfield Civic Centre, Enfield Borough, Middlesex, United Kingdom, in 1999. Stainless steel exterior cladding is type 316 with either a bright annealed or an embossed finish. Courtesy of the Nickel Development Institute (1999, Catherine Houska)

Fig. H Pier Pavilion, Herne Bay, Kent, United Kingdom, in 1999. This two- and three-story sports center on the end of an ocean pier has corrugated external type 316 stainless steel cladding. Courtesy of the Nickel Development Institute (1999, Catherine Houska)

Fig. I ICI Building, North York, Ontario, Canada, in 1981. Type 304 stainless steel panel for the curtain wall and four revolving door entrances

Fig. J Sun Life Centre, Toronto, Ontario, Canada, in 1999. Curtain wall is type 304 stainless steel face panels with a thickness of 0.06 inches (1.5 millimeters). Courtesy of the Nickel Development Institute (1999, Catherine Houska)

Fig. K Michael Fowler Centre, Wellington, New Zealand, in 2000. Curved stainless steel panels cover the top and bottom of the protruding circular center of the building. There are also vertical stainless steel sunshades between the windows. Courtesy of the Nickel Development Institute (1999, Catherine Houska)

Fig. L Statue of Genghis Khan fabricated of 250 tons of stainless steel. Located 54 kilometers from Ulaanbaatar, Mongolia, this statue is 131 feet tall.

Fig. M Cloud Gate sculpture by Anish Kapoor in Millennium Park, Chicago, Illinois.

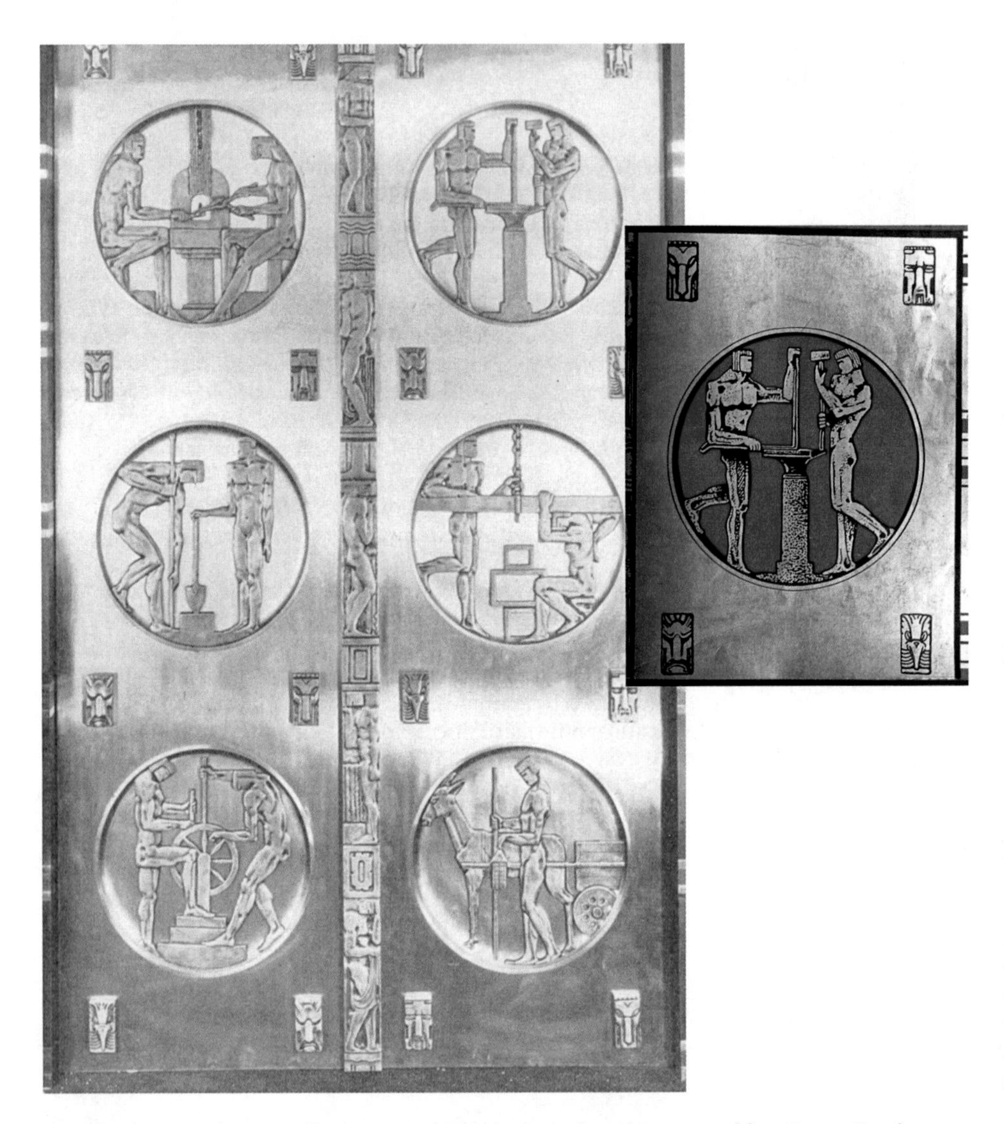

Fig. N Stainless steel entrance door designed and executed by Oscar Bach. Plaques represent industries of mining, smelting, fabrication, machining, building, and transportation. Top right plaque in color represents fabrication. Source: *Metal Progress,* June 1936, p 36

Fig. O Dramatic façade of Federal Savings and Loan Association in Brookfield, Ill. Exterior columns are sheathed in stainless steel, and the building is accented by stainless steel mullions. Source: *New Horizons in Architecture with Stainless Steel,* American Iron and Steel Institute, 1965, with permission

Fig. P Escalator, referred to as a "new electric stairway," uses bright stainless steel stringers, railings, and posts. Sears, Roebuck and Company store in Atlanta, Georgia. Source: *New Horizons in Architecture with Stainless Steel,* American Iron and Steel Institute, 1965, with permission

Fig. Q Stainless steel stairwell with single stringer and cantilevered steps. Architects: Daniel Badani, Michael Folliasson, Abro Kandjian, and Pierre Roux-Dorlut. Photo: Baltazar Korab. Source: *New Horizons in Architecture with Stainless Steel,* American Iron and Steel Institute, 1965, with permission

In an interview published in the June 1936 issue of *Metal Progress*, Oscar Bach described his impressions of stainless steel as an important new medium:

> "About a decade ago a new metal or a new combination of metals was created which I consider the most remarkable material ever made available for use in exterior and interior architectural facades. It is stainless steel.
>
> "Other materials—stone, marble, terra cotta, tile, bronze, iron, lead and zinc—have left their marks on respective periods in our civilization. So will this new material set its everlasting mark on our century and future civilization.
>
> "Just as we look back at the old times and modes of construction with their stone walls and wonder at the progress that has since been made in steel skeleton building construction and its mechanical wonders, so will generations to come look back upon the construction of the coming era and marvel at the innovations that stainless steel made possible.
>
> "Stainless steels bring to the arts and sciences an entirely new medium of expression—whether for artistic purposes or for the almost limitless utilitarian uses where a permanent metal is needed."

Mr. Bach then continued:

> "Early sculptors expressed themselves in three dimensions through the use of marble, granite and other materials such as the softer metals—all that were then available. The natural characteristics of these materials created limitations in their use. For decorative purposes, most large sculpturing, even when in low relief, necessitates a tremendous amount of bracing and support, as the material used can seldom carry any structural load. Now the stainless steels make it possible for artists to work in a material that is as strong and dependable as it is beautiful."

The beauty and strengths of stainless steel were indeed recognized.

Aircraft

Aircraft were one of the very earliest applications of stainless steels. In fact, in 1914, just one year after Harry Brearley's discovery of chromium stainless steel for cutlery, the British Munitions Board requisitioned all of this steel that could be made, because it was discovered that Brearley's steel would make excellent airplane engine exhaust valves. During the war, Thomas Firth & Sons sold their entire production of the steel under the name Firth's Aeroplane Steel, or FAS, for the valves in British fighter planes.

Stainless Steel Airplane—The Pioneer. In 1930, the largest manufacturer of steel automobile bodies in America, the Edward G. Budd Manufacturing Company of Philadelphia, was about to close its doors. It was the Great Depression, and automobile sales had plummeted. Mr. Budd, however, was not one to wait and see when business may pick up. He was intrigued with the new stainless steel that had just been introduced to America on the top of the tallest building in the world, the Chrysler Building in New York City.

Budd knew how to form large sheets of steel into intricate shapes and weld them together to create the strongest possible structures. He also learned that the new 18-8 stainless steel could be produced as very thin sheets with a tensile strength three times that of structural steel. He visualized the possibility of building a very strong and light structure, one that had never been built before, because no one in the 20 years since the metal had been developed had found a way of welding the high-strength metal together without softening it at the welds or destroying the corrosion resistance at the welded area.

The Budd Company had been spot welding automobile bodies for years, and they were experts in that business. They thought they could become the experts on welding stainless steel. Budd's chief engineer, Colonel Earl Ragsdale, and his assistants worked on that difficult welding problem and, in time, solved it. Pieces of the metal could be overlapped and welded together without diminishing the very high strength, and tests confirmed that there was no damage to the corrosion resistance.

Mr. Budd now wanted to produce a device or machine that would capture attention and show the kind of work that his company was capable of doing. What he decided to build was a flying machine. Planes were new in those days, and people were thrilled to see them. Few had flown in one. Budd's people thought he had gone mad and would be wasting money, even though it was money from his private account, on such a foolhardy project with no intention of getting into the aircraft business. However, Budd was something of a showman, and that's what he wanted to embark on—a foolhardy mission. Did someone have a better idea?

Within 18 months, a shining, stainless steel amphibian biplane with twin engines was being rolled out of the Budd factory on Hunting Park Avenue in Philadelphia. The plane had four seats in an open cockpit. It would have a cruising speed of 85 miles per hour and a cruising range of over 400 miles. The plane had been built from the plans of an Italian plane, substituting stainless steel for wood. The wings were covered with fabric.

The plane, named the Pioneer I, did get off the ground, and demonstration flights were made in America and Europe. After 1000 hours of flight, the Pioneer was brought back to Philadelphia. Mr. Budd donated the plane to the Franklin Institute in Philadelphia, where the plane is still on display in front of the building.

The RB-1 Conestoga Cargo Plane. During World War II, the Budd Company in Philadelphia received contracts from the Army and the Navy to build 600 cargo planes almost entirely of stainless steel. The contracts were placed because there were shortages of cargo planes and aluminum.

The Budd Company designed the RB-1 using an innovative ramp beneath the tail that lowered for easy loading of trucks and other cargo. Twenty-five planes were built by the time the war ended, and the balance of the contracts was cancelled. Seventeen of the planes were purchased by the Flying Tiger Air Freight Company and became their entire fleet for several years. These were the only cargo planes ever built of stainless steel.

Jet Engines. Frank Whittle received a patent in England for a gas turbine engine of the type used for modern aircraft. Airplanes with jet engines became widely used following World War II. Stainless steel is used in the engines for compressor blades and vanes and for the cylindrical compressor housings.

The Concorde was the only commercial supersonic aircraft. It was produced by the combined manufacturing efforts of Aérospatiale of France and the British Aircraft Corporation. Twenty of the aircraft were built and first placed into service in 1976. The planes flew at an average speed of Mach 2.02, or approximately 1330 miles per hour. The high speed created so much heat at the outer surface of the plane that the rudder and ailerons were required to be made of stainless steel rather than aluminum. The planes were in service for 27 years.

Automobiles

Ford Motor Company. The first use of stainless steel in America seems to have been by Henry Ford on models built in 1928 and 1929. The applications were for trim, radiator grilles, and hubcaps that were normally made of chromium-plated steel, which was often not of top quality. The chromium plate frequently developed pits in a few years. For the most part, cars were built to become obsolete in a few years. The manufacturers were very cost-conscious. Stainless steel, however, looked attractive and would last for the life of the car.

Rolls-Royce Motor Cars in England came out with a striking stainless steel radiator grille and cap in 1929, with the intent of adding the finest possible decorative touch to their stately cars.

The First Stainless Steel Automobiles. In 1935, officials at Allegheny Ludlum and the Ford Motor Company collaborated on an experiment intended to be an unusual advertisement for stainless steel. Six Ford Deluxe sedans (Fig. 53) were manufactured for display by Allegheny Ludlum to demonstrate the formability of 18-8 stainless steel and to show its beauty. The cars were driven by Allegheny officials to important events and when visiting important customers. It was apparently never the intent to suggest that automakers should start making their cars of stainless steel, although DeLorean did just that in 1981.

The DeLorean All-Stainless Steel Car. John DeLorean (1925–2005), the youngest executive at General Motors, was well known in the automobile industry as a capable engineer. He had the idea that he could build a futuristic sports car that would attract a lot of buyers.

Fig. 53 Ford Tudor stainless sedan

With this thought in mind, he left General Motors in 1975 and founded the DeLorean Motor Company. He was in the prime of life at the age of 50.

DeLorean decided to build a factory in a country where unemployment was especially high as a way to find the cheapest possible labor. Ireland's Industrial Development Board offered him certain incentives, which he accepted.

Construction of his plant in Dumurry, a suburb of Lisburn, began in October 1978. The DeLorean Motor Car, Ltd. plant was completed in just 18 months. The plant was actually on a line between two communities: Twinbrook, which was a Catholic town, and Dumurry, which was Protestant. There was an entrance from each community, but it is said that this was more of a convenience than for religious segregation.

In early 1981, the assembly line started, albeit with few skilled workers and apparently not a lot of training. Meanwhile, DeLorean dealerships were being lined up exclusively in the United States, in New Jersey, Michigan, and California. The futuristic, streamlined car with its gleaming stainless steel body and gull-wing doors was to sell for $25,000 ($57,500 in 2008 dollars), and there was an impressive waiting list of anxious customers. These were the only cars made in production of all-stainless steel bodies. This and the outlandish design were the main attractions. They were two-door cars. The so-called gull-wing doors were of a design never seen before or since. The doors opened from the bottom up and remained high in the air while the passengers stepped in or out.

Every car looked identical. There was no painting, although it is said that some dealers offered to paint a car if a customer insisted. Approximately 9000 cars had been built by December 1982, but sales were not up to expectations, and the firm went bankrupt.

Meanwhile, sometime in 1982, John DeLorean became the target of a Federal Bureau of Investigation sting operation designed to arrest drug-trafficking individuals. DeLorean was acquitted of all charges, but his reputation was forever tarnished.

Of the 9000 DeLoreans produced, it is estimated that there are still approximately 6000 on the road. There is a strong affinity among DeLorean owners and their clubs.

Trains

Steam Locomotives. J.H.G. Monypenny wrote, "The present year (1925) has been notable in the engineering world as being a railway centenary; it may not be amiss, therefore, if the locomotive is taken as an example which offers suitable opportunities for testing the properties of stainless steel in connection with steam service."

In approximately 1922, steam locomotives in England began to be fitted with the cutlery type of stainless steel for many types of valves, including safety valves, steam whistle valves, steam brake valves, and pressure gage valves. The valves were said to last far longer than when made of traditional metals, such as gun metal and other copper alloys.

Passenger and Commuter Trains. The first stainless steel trains by the Budd Company in 1934 revolutionized the way passenger trains were built (Chapter 8, "Edward G. Budd (1870–1946), Inventor and Entrepreneur"). Upon expiration of the Budd patents, stainless steel trains were built in many countries and represent the prevailing type of construction for trains and subway cars to the present time.

Utilitarian Low-Cost Stainless for Coal Hopper Cars. A chromium content of 10.5% is the lowest amount of chromium required for a steel to be called stainless. Although alloy 3CR12 (currently listed as UNS S41003) actually does rust, it also forms a tightly adherent oxide coating. The first 25 ton trial heat of the 3CR12 ferritic-martensitic chromium stainless steel was cast at Middleburg Steel and Alloys, Middleburg, South Africa. This utilitarian low-cost stainless would be widely used for applications not requiring a shiny finish, such as for coal cars.

CHAPTER 13

Canada Restores a Fleet of Stainless Steel Railcars

"There may be no better testimonial to the quality and longevity of stainless steel railcars than the award-winning modernization of Canada's transcontinental passenger fleet."
James Borland, 1995

THE RESTORATION of a fleet of trains had never before been contemplated, let alone attempted, but VIA Rail, the publicly owned Canadian passenger train company, decided to refurbish its 40-year-old fleet of stainless steel railcars. The cars still looked fine. There were no dents, and there was no rusting. The spot welds made by the patented Budd process, identifiable by their smooth brown spots, had never been known to fail by metal fatigue. When washed with soap and water, the cars looked as good as new.

VIA Rail estimated that the cost of restoration would be less than $1 million per car, which they figured was probably about half of what it would cost for a new car. In other words, it could be said that each stainless steel shell is worth approximately $1 million.

The restoration work was done at the AMF Technotransport, Inc. shops in Montreal. The program originally called for the upgrading of 181 stainless steel railcars. The E.G. Budd Manufacturing Company of Philadelphia had sold 157 of the cars to the Canadian Pacific Railway in 1955 to 1956. The remaining cars of approximately the same age were supplied by other companies. "They were reliable, those cars," said Roger Hoather, VIA's Project Manager. "It (stainless steel) is perceived as being a luxury, but it really is a very practical material."

Everything inside and outside the cars was replaced, including the undercarriages, wheels, motors, electric generators, and braking systems. The interior furnishings and passenger amenities were as fine as could be found on a luxury cruise ship. It was a tremendous undertaking that cost approximately $181 million. Never had any company invested so much money in restoring second-hand equipment.

The cars are even lighter than when originally built, because of the elimination of individual generators for heating, cooling, and lighting in each car, but still approximately 15% heavier than cars made today with computer-assisted designs that reduce section thickness of the stainless steel frame without sacrificing structural integrity. (It is interesting to consider that Albert Dean, the young aeronautical engineer, had designed the Budd trains within only 15% of the ideal weight by using mathematical equations and a slide rule.)

The original frame of the cars was type 201 stainless steel, which was one of the earliest uses of the higher-manganese, lower-nickel alloys introduced by Allegheny Ludlum to conserve nickel. The alloy was found to have properties equivalent to type 301 alloy. Minor repairs to the framing were made with type 301L alloy, and any repairs to the skin were made with type 304. The restored trains were first put into service in 1992. The 181 cars would make a train three miles long.

The restored cars were chosen to receive the international Brunel Award in 1994 by a panel of experts from the Foundation for Railway Transportation Excellence, the U.S. Federal Railway Administration, Amtrak, and the Watford Group of European Architects and Designers.

Hoather, of VIA Rail, said the cars could probably be used for another 40 years, assuming two or three interior redecorations, but officially, the program is only supposed to extend the life of the cars by 20 years.

The VIA restoration program was so successful that a second restoration project was immediately initiated. Trains were hauled out of graveyards and removed from unused sidings all over North America in a frantic effort to salvage as many stainless steel cars as possible. It was now realized that these cars were like hidden treasure, with each one being worth approximately $1 million.

The searchers were somewhat disappointed when they found only 40 cars. They accepted one with bullet holes, and there was also a circus train that had belonged to the Ringling Brothers and Barnum and Bailey Circus.

CHAPTER 14

The Plummer Classification System of Trade Names

BY 1934, the martensitic chromium type of stainless steel had been made in America for 19 years. The ferritic chromium type had been made in America for approximately 14 years, and the chromium-nickel steels had only been introduced to America for 7 years. Approximately 100 companies in the United States and Canada had entered the stainless steel business, selling their steel under trade names that often did not indicate in any way their essential compositions.

The numbering systems of the American Iron and Steel Institute (AISI) and the Alloy Casting Institute (ACI) had recently been set up, but most stainless steel producers preferred to use only their trade names. The situation was puzzling. Many customers were contacted by salesmen who recommended alloys that may or may not have been different from alloys promoted by a competitor. Nor were there any published lists of trade names to sort this out.

In 1931, Clayton L. Plummer, Technical Director of the Robert W. Hunt Company, undertook the preparation of a classification system. The R.W. Hunt Company of Chicago was the world's largest firm of consulting, testing, and inspecting engineers and had a special need for some kind of classification system of stainless steels. Plummer published his 22-page classification in *Factory & Industrial Management* in 1931. He developed his system using data on chemical compositions obtained from each of the manufacturers and divided stainless steels into 39 groups.

However, Plummer's original classification did not actually include the chemical ranges for the groupings, but instead, a code number was used. Because this arrangement proved to be confusing, Plummer revised and updated his classification in 1934 and included the actual chemical ranges or limits for each grouping. He listed the chromium alloys according to different ranges of chromium and different ranges of carbon within each chromium range. The first group, for example had alloys with a chromium range of 5 to 7% and 0.12% maximum carbon. The next group had alloys with 5 to 7% chromium and 0.13 to 0.20% carbon, and a third grouping had alloys with 5 to 7% chromium and 0.13 to 0.20% carbon but unspecified amounts of molybdenum.

One classification of Plummer's system is shown in Table 3. It is for alloys with 16 to 23% chromium, 7 to 11% nickel, and 0.13 to 0.20% carbon. The example shows the trade names and the producer of each alloy. The Plummer system was unwieldy and was replaced by the three-digit AISI numbering system.

Table 3 Example of the Plummer classification system for alloys with 16 to 23% chromium, 7 to 11% nickel, and 0.13 to 0.20% carbon

Code	Trade name	Company
100	Acme Silchrome KA-2	Acme Steel Co., Chicago
105	Allegheny Metal	Allegheny Steel Co., Brackenridge, Pa.
110	SS	Alloy Metal Wire, Moore, Pa.
112	AISI Std. 302	American Iron and Steel Institute, New York, N.Y.
118	Armco 18-8	American Rolling Mill, Middletown, Ohio
120	USS 18-8	United States Steel Corp., Pittsburgh, Pa.
135	B & W No. 600	Babcock & Wilcox Company, New York, N.Y.
145	Bethadur No. 2	Bethlehem Steel Co., Bethlehem, Pa.
158	Anka H	Brown Bailey's Steel Works, Montreal
165	Calite E	The Calorizing Company, Pittsburgh, Pa.
170	No-Kor-O	Canadian Atlas Steels, Ltd., Welland, Ontario
185	KA-2	Chrome Alloy Products, Inc., Nicetown, Philadelphia, Pa.
195	Vasco Stainless U	Colonial Steel Co., Pittsburgh, Pa.
205	Sweetaloy 17 (KA-2)	Cooper Alloy Foundry, Elizabeth, N.J.
210	Rezistal RA-2H	Crucible Steel Casting Co., Cleveland, Ohio
215	Rezistal RA-2	Crucible Steel Casting Co., Cleveland, Ohio
220	Uniloy 18-8	Cyclops Steel Company, Titusville, Pa.
240	DH KA-2	Driver-Harris Co., Harrison, N.J.
245	Duraloy 18-8	Duraloy Co., Pittsburgh, Pa.
250	Durco KA-2	The Duiron Company, Dayton, Ohio
255	Esco Stainless No. 40	Electric Steel Foundry Co., Portland, Ore.
260	Thermalloy 18-8	Electro Alloys Co., Elyria, Ohio
265	Empire 18-8	Empire Steel Castings Co., Reading, Pa.
268	Fahralloy F-8	Fahralloy Co., Chicago
275	Sterling Nirosta FC	Firth-Sterling Steel Co., McKeesport, Pa.
285	Q-Alloy Chrome KA-2	General Alloys Co., Boston, Mass.
295	Rezistal KA-2	Halcomb Steel Co., Syracuse, N.Y.
315	Industrial No. 5188	Industrial Steels Co.
335	Corrosion Resistant R-3	Wm. Jessop & Sons, Inc., New York City
340	Hi Gloss	Jessop Steel Co., Washington, Pa.
355	Stainless Iron	Latrobe Electric Steel Co., Latrobe, Pa.
360	Corrosion Resistant Circle L No. 23	Lebanon Steel Foundry, Lebanon, Pa.
363	Silchrome KA-2	Ludlum Steel Co., Watervliet, N.Y.
365	Nirosta 18-8	Michiana Products Co., Michigan City, Ind.
370	Misco 18-8	Michigan Steel Castings Co., Detroit, Mich.
375	Midvaloy 18-08	Midvale Steel Co., Nicetown, Philadelphia, Pa.
383	Nationalloy Grade 1	National Alloy Steel Co., Blawnox, Pa.
400	Fahrite Grade –2	Ohio Steel Foundry Co., Springfield, Ohio
420	Enduro 18-8 (type 302)	Republic Steel Corp., Youngstown, Ohio
420	Enduro 19-9 (type 305)	Republic Steel Corp., Youngstown, Ohio
425	Defistain	Rustless Iron Corp. of America, Baltimore, Md.
430	Sharon Stainless 18-8	Sharon Steel Hoop Co., Sharon, Pa.
435	Shawinigan Nirosta	Shawinigan Chemicals, Ltd., Montreal
445	Standard Alloy HR No. 7	Standard Alloy Co., Cleveland, Ohio
447	Stanley Grade D	The Stanley Works, New Britain, Conn.
450	Tisco 101	Taylor-Wharton Iron & Steel Co., High Bridge, N.J.
450	Tisco 103	Taylor-Wharton Iron & Steel Co., High Bridge, N.J.
450	Tisco 110	Taylor-Wharton Iron & Steel Co., High Bridge, N.J.
465	Timken 18-8	Timken Steel & Tube Co., Canton, Ohio
470	UHB Stainless 3H	Uddeholm Co. of America, New York City
485	Warman 18-8	Warman Steel Castings Co., Los Angeles, Calif.
495	Wehralloy No. 22	Wehr Steel Co., Milwaukee, Wis.
500	Whelco No. 2	Wheelock & Lovejoy Co., Cambridge, Mass.

The History of Stainless Steel
Harold M. Cobb

CHAPTER 15

The Unified Numbering System (UNS) for Metals and Alloys

IN APPROXIMATELY 1950, the U.S. Department of Defense (DoD) found a need to produce the first catalog of all metals and alloys cited in the specifications of the various and sundry standards organizations in the private sector, such as the Society of Automotive Engineers (SAE) and the American Society for Testing and Materials (ASTM), and in the governmental specifications, such as the Federal QQ Standards and Military MIL standards. The DoD published this information in MIL-HDBK-H1, *Cross Index of Chemically Equivalent Specifications and Identification Code (Ferrous and Nonferrous)*. The book contained the chemical compositions of each of the alloys listed.

MIL-HDBK-H1 was first published in the early 1950s as a typewritten book of approximately 200 pages. It was thereafter the practice of the DoD to solicit help from the various standards organizations in updating the book.

It was found that the same alloys were sometimes listed with different designations and slightly different compositions by the various standards organizations. The problems were reduced for those alloys that were listed according to the numbers of the Aluminum Association, the Copper Development Association, and the American Iron and Steel Institute (AISI).

The designations for nickel alloys were a particular problem. There had never been a numbering system for these alloys. Because trade names were not accepted in ASTM specifications, designations were resorted to, such as "Nickel-Chromium-Molybdenum-Iron-Cobalt Alloy" and "Nickel-Cobalt-Chromium-Molybdenum-Titanium-Aluminum." These names were unsatisfactory to all concerned.

There were other problems with metal designations. The Copper Development Association and the American Welding Society had numbering systems, but they were outmoded. One of the biggest problems was that AISI announced early in the 1960s that they were discontinuing the practice of assigning numbers to steels.

To take care of an immediate need for numbers for proprietary steels in ASTM specifications, a makeshift system of XM numbers, from XM-1 to XM-24, was established. Another group within ASTM started using a 600 series of numbers that looked like AISI numbers but had never been officially recognized by that organization. The standards system for metals was in disarray.

Because of these problems, in 1967 ASTM and SAE began to explore the possibility of developing a new numbering system for metals that would be applicable to all alloys. The U.S. Army was especially interested in this subject, and in May 1969, a contract was issued by the Army Materials and Mechanics Research Center to SAE to conduct a "Feasibility Study of a Unified Numbering System for Metals and Alloys." This project was jointly sponsored by ASTM and SAE, and a committee was appointed to conduct the study. The committee was chaired by Norman L. Mochel, a past president of ASTM, and consisted of the following members:

- Herbert F. Campbell, Army Materials and Mechanics Research Center
- Harold M. Cobb, ASTM
- Alvin G. Cook, Allegheny Steel Corp.
- Henry B. Fernald, technical consultant
- Muir L. Frey, engineering consultant
- S.T. Main, Grumman Aircraft Corp.
- Norman L. Mochel, chairman
- R. Thomas Northrup, SAE
- Bruce A. Smith, General Motors engineering staff
- Harry H. Stout, Phelps Dodge Copper Products Corp.

Many individuals were contacted during the course of the 18 month feasibility study, as well as the major trade associations concerned with metals numbering systems, including The Aluminum Association, AISI, and the Steel Founders' Society of America (SFSA). It was recognized at the outset that any new system could be successful only if these organizations were in general agreement with the concept. In January 1971, the study was completed, and a report was submitted to the Army, stating that it had been determined that a unified numbering system for metals was feasible and desirable. The report included a general proposal of how such a system could be established to provide a coherent designation system for all current and future metals and alloys.

In April 1972, ASTM and SAE established an Advisory Board to further develop and refine the proposed numbering system. The Advisory Board consisted of the following members:

- Chairman: Bruce A. Smith, General Motors engineering staff
- Secretaries: Harold M. Cobb, ASTM staff, and R. Thomas Northrup, SAE staff
- John Artman, Defense Industrial Supply Center
- Lawrence H. Bennett, National Bureau of Standards
- Alvin G. Cook, Allegheny Ludlum Steel Corp.
- Henry B. Fernald, Jr., technical consultant
- John Gadbut, International Nickel Co.
- Joseph M. Engel, Republic Steel Corp. (representing AISI)
- W. Stuart Lyman, Copper Development Association
- Robert E. Lyons, Federal Supply Service
- Norman L. Mochel, metallurgical consultant
- Edward F. Parker, General Electric Co.
- Richard R. Senz, The Aluminum Co. of America (representing the Aluminum Association)
- Whitney Snyder, American Motors Corp.
- Rudolph Zillman, SFSA

This board decided that the official name of the new system would be the Unified Numbering System (UNS) for Metals and Alloys and that the guiding principles of the system would be as follows:

- Each designation for a metal or alloy should pertain to a specific metal or alloy as determined by its unique chemical composition or to its mechanical properties or physical characteristics when these are the primary defining criteria and the chemical composition is secondary or not significant.
- For ease of recognition, the numbers assigned should incorporate numbers from existing numbering systems whenever possible.

- The numbering system should be designed to accommodate current metals and alloys and to anticipate the need to provide numbers for new alloys for the foreseeable future.

In March 1974, the UNS Advisory Board completed the "SAE/ASTM Recommended Practice for Numbering Metals and Alloys," which established 18 series of designations that consisted of a prefix letter and five digits, as shown in Table 4. It should be noted that, in most cases, the letter is suggestive of the family of metals identified.

The procedure for assigning numbers to each of the 18 series of numbers was to be coordinated by the Advisory Board, but the specific details for each series would be developed by experts in each of those fields. Examples of some of these designations are shown in Table 5.

By 1974, the Advisory Board had coordinated the establishment of specific UNS designations for over 1000 metals and alloys, including steel, stainless steel, tool steel, superalloys, aluminum, copper, cobalt, magnesium, and nickel. These designations were listed in the first edition of the *UNS Handbook*. Each entry included the UNS designation, a brief description of the alloy, the chemical composition of the alloy, and a list of the national specifications in which the alloy appeared.

Alvin G. Cook of the Allegheny Steel Corporation agreed to serve as chairman of the Advisory Board of the Unified System for Metals and Alloys. Cook was a well-known metallurgist who served on many ASTM and SAE committees. His job was to appoint individuals who would be responsible for establishing a numbering system suitable for each of the 18 series of numbers and for assigning numbers to each alloy within that category. For aluminum, copper, and the AISI steel,

Table 4 Primary series of Unified Numbering System (UNS) numbers

UNS series	Metal	UNS series	Metal
A*xxxxx*	Aluminum and aluminum alloys	L*xxxxx*	Low-melting metals and alloys
C*xxxxx*	Copper and copper alloys	M*xxxxx*	Miscellaneous nonferrous metals and alloys
D*xxxxx*	Steels—designated by mechanical property	N*xxxxx*	Nickel and nickel alloys
E*xxxxx*	Rare earth and rare-earth-like alloys	P*xxxxx*	Precious metals and alloys
F*xxxxx*	Cast irons	R*xxxxx*	Reactive and refractory metals and alloys
G*xxxxx*	AISI and SAE carbon and alloy steels	S*xxxxx*	Stainless steels, valve steels, and superalloys
H*xxxxx*	AISI H-steels	T*xxxxx*	Tool steels
J*xxxxx*	Cast steels	W*xxxxx*	Welding filler metals and electrodes
K*xxxxx*	Miscellaneous steels and ferrous alloys	Z*xxxxx*	Zinc and zinc alloys

Table 5 Examples of Unified Numbering System (UNS) designations

UNS No.	Traditional designation
A03190	AA 319.0 (aluminum alloy casting)
A92024	AA 2024 (wrought aluminum alloy)
C26200	CDA 262 (cartridge brass)
G12144	AISI 12L14 (leaded-alloy steel)
G41300	AISI 4130 (alloy steel)
K93600	Invar (36% nickel alloy steel)
L13700	Alloy Sn 70 (tin-lead solder)
N06007	Nickel-chromium alloy (Hastelloy G)
N06625	Alloy 625 (nickel-chromium-molybdenum-niobium alloy)
R58210	Alloy 21 (titanium alloy)
S30452	AISI 304N (stainless steel, high nitrogen)
T30108	AISI A-8 (tool steel)
W30710	AWS E307 (stainless steel electrode)
Z33520	Alloy AG40A (zinc alloy)

it would be the incorporation of those numbers within the UNS system, such as A20240 for aluminum alloy 2024.

The Stainless Steel Numbering System

Cook would be responsible for the S-series for wrought stainless steels. For modifications of the original three-digit AISI numbers for stainless steel, Cook set up a complex but logical system assigning code numbers to the last two digits, with "00" being the code for the basic alloy.

The code for 0.03 maximum carbon would be "03," with type 304L becoming S30403. Code numbers were set up to indicate additions to the basic alloy, including "10 to 14" for manganese, "15 to 19" for silicon, "20 to 24" for sulfur or selenium, "25 to 29" for molybdenum, "30 to 34" for copper, "35 to 39" for titanium, "40 to 45" for columbium (niobium), "45 to 49" for aluminum, "50 to 59" for nitrogen, "60 to 69" for boron, and "80 to 89" for filler metals. Type 304LN would become S30453.

With the addition of new stainless alloys, the plan was to incorporate the original designation when possible. Thus, AM 350 would become S35000, and alloy A286 would become S66286.

Miscellaneous Steels. For miscellaneous steels and ferrous metals, a K-series numbering system was devised by Harold M. Cobb of the ASTM staff. No standard numbers had ever been established for these materials. With the number of available digits limited to five, it was not possible to develop a code based on the alloying elements. Instead, a system was devised wherein the UNS designation would indicate the total alloy content and the carbon content.

Metals with total alloy contents of 9% or more fall into the K9*xxxx* series. In this system, the 36% nickel steel called Invar has the UNS number K93600, where the second, third, and fourth digits, "360," indicate that the total alloy content is 36.0%.

Carbon, Alloy, and Stainless Steel Castings. A J-series was set up for numbering carbon, alloy, and stainless steel castings. There had been no general numbering system for carbon and alloy castings, but the Alloy Casting Institute (ACI) had developed a widely used alphanumeric system for corrosion- and heat-resistant steels.

The ACI numbering system developed in the 1930s was merged into the activities of the SFSA and later came to be managed by ASTM Subcommittee A01.18 on Steel Castings. The ACI system was codified to give a general indication of the chemical composition of an alloy. It was not an all-digit system that could be incorporated into the UNS system.

The SFSA decided to adopt the same system for designating steel castings as that developed by Cobb for the K-series of miscellaneous steels, wherein the alloys are codified in accordance with their total alloy content. For example, ACI designation CA40, which is similar to type 420, has a chromium content of 11.5 to 14.0%. The lower number of the chromium range, 11.5, is used for the total alloy content and inserted as the second, third, and fourth digits, so that the UNS designation becomes J91150.

Welding Alloys. For welding alloys, a comprehensive system was set up by R. David Thomas, Jr., representing the American Welding Society. Thomas was president of Arcos, a welding filler metal company in Philadelphia. The UNS W-series was set aside for ferrous and nonferrous welding alloys. Welding filler metals that are similar to the basic stainless steels in the S-series would be included in the S-series, with the numbers 80 to 89 in the last two digits. Thus, S30480 is for a welding filler metal similar to type 304.

Composite welding alloys are to be included in the W-series as W300*xx* for type 300 alloys and W400*xx* for type 400 alloys.

UNS Advisory Committee. In the early 1990s, Alvin Cook resigned as the chairman of the Advisory Committee, having served since 1974. James D. Redmond of TMR Stainless, Pittsburgh, Pennsylvania, assumed that position, including the responsibility for assigning numbers to stainless steels.

CHAPTER 16

The Naming and Numbering of Stainless Steels

FOR MANY YEARS, the term *stainless steel* was applied only to those alloys that were hardenable by heat treatment, the martensitic alloys. This was logical, because the martensitic alloys had all of the characteristics of alloy steel, except that they did not rust. Eventually, however, several different classes of steels were recognized as having sufficient chromium content for stain-resisting alloys. One early publication that paved the way for the acceptance of stainless steel as a generic name for these high-chromium steels was *The Book of Stainless Steels*, published by the American Society for Metals in 1935. Today, the term *stainless steel* in other languages includes:

Czech	Nerezová ocel
Dutch	In roestvast staal
French	Acier inoxidable
German	Nichtrostende Stähle
Italian	Acciaio inossidabile
Polish	Gatunków stali
Portuguese	Aço inox
Russian	Nerzhaveiushchei stali
Spanish	Acero inoxidable
Swedish	Rostfritt stål

Early Classes of Stainless Steel

The English term *stainless steel* appeared from the commercial development of cutlery steels in 1912 by Harry Brearley with the Brown Firth Research Laboratories in Sheffield, England. In the same year, austenitic chromium-nickel steel was developed and studied by Eduard Maurer and Benno Strauss at the Krupp Laboratories in Essen, Germany. Unlike the hardenable martensitic chromium steel of Brearley, the Krupp alloy steel had a softer austenitic crystal structure (due to the higher nickel content of the Krupp alloy). This softer crystal structure has important benefits for fabricating shapes. The third class of steels, which would become known as stainless iron or rustless iron, was discovered by Christian Dantsizen of the General Electric Research Laboratory at Schenectady, New York, in 1911. Although the third class was later to be recognized as ferritic stainless steels, the term *rustless iron* was initially used, because the steel was not hardenable like martensitic steel.

The Martensitic Alloys. The Brown Firth alloy produced in 1912 was an alloy steel with approximately 13% chromium and 0.24% carbon, a composition falling within the limits of what became known in America as type 420 stainless steel. By 1914, the alloy was being made into table knives by George Ibberson & Co., a Sheffield cutler. The blades were stamped with the manufacturer's name and the words "Stainless Knife." This is the first known use of the word *stainless* to describe the alloy. Also in 1914, the steel was being sold for the production of aircraft engine valves under the name Firth's Aeroplane Steel (FAS). In fact, the entire output of FAS was destined to be used for this purpose throughout the war years of 1914 to 1918. In 1917, the Firth-Brearley Stainless Steel Syndicate was formed to foster the worldwide production of stainless cutlery steel.

The Austenitic Alloys. The Krupp alloy that was developed in 1912 was austenitic and had a chemical composition that was virtually the 18% chromium and 8% nickel composition of 18-8, or what became known as type 304 stainless steel. Krupp named the alloy V.2.A and referred to the material as a corrosion-resistant iron-chromium-nickel alloy. The alloy was a new metal very different from steel. It could not be hardened by heat treatment, it had a different crystal structure, it had vastly different physical properties, it remained ductile at subzero temperatures, it was nonmagnetic unless work hardened, and it did not rust. Krupp marketed the alloy under the trade name Nirosta

KA2, with Nirosta being an abbreviation of *Nichtrostende Stähle* (nonrusting steel).

The Ferritic Alloys. Dantsizen's alloy contained 14 to 16% chromium and 0.07 to 0.15% carbon. This ferritic alloy was only slightly hardenable by heat treatment and, at that time, was called stainless iron because of its relative softness and ductility. The alloy was initially used for leading-in wires for incandescent lamps. It is also interesting to note that there was once a Rustless Iron and Steel Company established in Baltimore in 1930. The classic book written by the English metallurgist J.H.G. Monypenny in 1926 was entitled *Stainless Iron and Steel*. The terms *stainless iron* and *rustless iron* referred to certain iron-chromium-carbon alloys that were predominantly ferritic and not hardenable by heat treatment or only slightly hardenable. However, the terms were never adequately defined and are no longer in general use.

Stainless Steel Trade Names

Some of the earliest trade names, from approximately 1915, for various stainless steel alloys were Anka (Brown Bayley's Steel Works, Ltd.), Staybrite (Thomas Firth & Sons, Ltd.), V.2.A. (Krupp Works), Jessop Hi-Gloss (Jessop Steel Co.), and Sterling Nirosta KA-2 S (Firth-Sterling Steel Co.).

By 1923, the Firth-Sterling Steel Company at McKeesport, Pa., was marketing three types of chromium stainless steels, which were called Stainless Steel Annealed, Stainless Steel Particularly Annealed, and Stainless Iron. They were "sold on brand and performance and not to carbon content or chemical analysis." Four years later, Firth-Sterling changed its marketing practice, offering five types: A, T, B, M, and H. For each type, the carbon and chromium contents were listed. Type A, for example, was listed as having 0.35% carbon and 13.5% chromium, a composition that meets the requirements of what became generally known as type 420 stainless steel.

By the early 1930s, there were almost 100 companies in the United States that produced or fabricated stainless steels, marketing them with almost 1000 trade names, such as Enduro S (Republic Steel), Allegheny Metal B (Allegheny Steel Co.), Industrial No. 35 (Industrial Steels Co.), Ship Brand Orange Label (Webb Wire Works), and Corrosion Resisting Circle L (Lebanon Steel Foundry). However, some companies broke away from this practice of giving their products mysterious names and began using meaningful names, such as Armco 18-8, USS 19-9, and Enduro 20-10, where the numbers indicated the approximate percentages of chromium and nickel in the alloys. Although the American Iron and Steel Institute numbers had been developed, they were not yet in general use.

Modern producers continue to name their alloys in various ways, including the custom of using established standards numbers, such American Iron and Steel Institute (AISI), Unified Numbering System (UNS), Deutsches Institut für Normung (DIN), or Euronorm numbers (EN) following their company names or trade names, as in the case of AL 410S, AK Steel 41003, Cronifer 3718, and Outokumpu 4372.

Standardization

Standardized designation of alloys is done by various specification organizations, as briefly described in this section. By way of example, Table 6 lists comparable designations for type 304 stainless steel.

American Iron and Steel Institute (AISI). There was no standardization of the chemical compositions of alloys produced in the stainless steel industry until the early 1930s. The now-familiar four-digit numbering system for carbon and alloy steels had been developed by AISI. They decided on a three-digit numbering system for stainless steels. It is important to note, however, that AISI unfortunately classed the 4 to 6% chromium steels as stainless steels, because these alloys showed a resistance to sulfides four to ten times that of ordinary steel. Because of this classification, the dictionary definition of stainless steel was stated to be an "alloy of iron and at least four percent chromium."

The AISI numbering system for stainless steels consisted of a 300 series for chromium-nickel austenitic alloys, a 400 series for high-chromium ferritic and martensitic alloys, and a 500 series for 4 to 6% chromium alloys. Examples in each series include 304, 410, and 501. Forty-six AISI numbers and their corresponding chemical compositions were published in 1932, twenty-eight of which are still in use. Many new AISI numbers were added that were primarily modifications, using suffix letters, such as 304L, 304H, 304N, and 420F. AISI 329 was the only duplex (austenitic-ferritic) alloy to have been assigned an AISI number.

In the 1960s, the AISI established a 200 series of numbers for the high-manganese austenitic alloys, and a 600 series of numbers was es-

Table 6 Comparable designations for type 304 stainless steel

Country or group	Designations
Australia	304 S30400
Austria	X5CrNi 18-10 KKW
Brazil	E304
Bulgaria	0Ch18N10
Canada	304 S30400
China (People's Republic of China)	0Cr19Ni9 S30408
Czech Republic	240
Spain	X6CrNi 19-10 F.3504; E-304
European	EN 1.4301 X5CrNi 18-10
France	Z6CN18-09 304F01
Germany	1.4301 X5CrNi 18-10
Great Britain	En 58E 304S18 EN 1.4301
Hungary	AoX 7CrNi 18-9 KO 33
India	04Cr18Ni 11
International Standards Organization (ISO)	X5CrNi 18-10
Italy	X8CrNi 19-10
Japan	SUS 304
Korea	STS 304
Mexico	MT304
Poland	0H8N
Romania	T6NiCr180
Russia	08Ch18N10
United States	304 30304 S30400

tablished to cover proprietary alloys, such as Allegheny Ludlum's A286 alloy that was designated as grade 660 and Armco's 17-4PH that was designated as grade 630. Although these 600-numbers still appear in ASTM specifications and elsewhere, they were never officially adopted by AISI.

By 1957, there were 39 AISI numbers for stainless steels, including the 4 to 6% chromium types 501 and 502. These 39 alloys are those that have often been referred to as standard stainless steels, whereas all of the remaining alloys have been referred to as nonstandard stainless steels, simply because they were not included among those alloys having assigned AISI numbers.

In the mid-1960s, AISI discontinued the practice of assigning numbers, because of questions concerning the legality of a trade association being in the business of writing standards (i.e., establishing chemical composition limits for stainless steel alloys). However, the AISI numbering system is still the basic system used in the United States for designating alloys standardized by AISI prior to 1960. The American Iron and Steel Institute, however, requested that the use of "AISI" should be discontinued, but that was easier said than done. The AISI numbers are used as the primary stainless numbering system in many countries, including Canada, Mexico, Brazil, South Africa, India, Australia, New Zealand, and, in a modified form, in the British, French, Japanese, and Korean numbering systems. In addition, the AISI numbers are widely recognized throughout the world and are often referenced or adopted by stainless steel producers.

The Society of Automotive Engineers (SAE) introduced a modification of the AISI stainless steel numbers for use in SAE standards. The SAE embedded most, but not all, of the AISI three-digit stainless steel numbers into a five-digit system, wherein the two digits "30" preceded the AISI 300-series numbers, and "51" preceded the AISI three-digit 400-series numbers. Thus, AISI 304 became SAE 30303, and AISI 410 became SAE 51410. For the stainless steel casting alloys, the prefix "60" was to precede AISI designations with comparable corrosion-resisting compositions to the Alloy Casting Institute designations. Therefore, SAE 60410 was the designation for CA-15, 60420 for CA-40, 604304 for CF-8, and 60304L for CF-3. For heat-resisting casting, the prefix "70" was used, such as 70446 for alloy HC.

The SAE also developed a series of special "EV" designations, such as EV-4 and EV-13, for proprietary high-chromium, heat-resisting aircraft engine valve steels.

American Society for Testing and Materials (ASTM). In the 1960s, ASTM Committee A-10 on Iron-Chromium, Iron-Chromium-Nickel, and Related Alloys introduced a series of 34 "XM" (XM-1 through XM-34) numbers to designate certain proprietary stainless steel alloys. This scheme was adopted so that the proprietary alloys could be included in specifications without using trade names, which are generally not accepted in ASTM specifications. The precipitation-hardening alloys were given "XM" numbers, and Armco's 15-5 PH alloy, for example, was assigned the designation XM-12.

The Alloy Casting Institute (ACI), a former division of the Steel Founders' Society of America, developed a system for designating stainless and heat-resisting casting alloys. Casting designations within this system begin with either the letter "C," for corrosion-resistant alloys, or "H," for heat-resistant alloys. Typical examples are CA-15 and HK-40. This is the most common designation system for stainless steel casting alloys in the United States. Although the ACI organization no longer exists, the system is maintained by ASTM Subcommittee A01.18 on Steel Castings.

British Standards Institution (BSI). During World War II, the BSI introduced emergency numbers (En), which was the first standard numbering system for steels in Great Britain. Stainless steels were assigned En numbers 56 through 61, with letter suffixes. An alloy similar to AISI 304, for example, received the designation En 58E. The En numbers, which should not be confused with the modern Euronorm numbers (EN), were superseded in approximately 1960, but they continue to be used and are often seen in reference books.

In approximately 1960, the BSI adopted a new numbering system for stainless steels that incorporated the AISI system in most, but not all, cases. The first three digits in the British system are the AISI numbers, followed by the letter "S," for stainless, and two digits to indicate modifications. Therefore, 304S18 and 410S21 are examples of the BSI numbers for types 304 and 410 stainless steels. Exceptions include numbers such as 301S81, which is not equivalent to AISI 301 but is a designation for the 17-7PH alloy, and 284S16, which has a composition similar to that of alloy 202.

There are also British Aerospace numbers for stainless steels. These include "HR" numbers, such as HR 3, "S" numbers, such as 2S.130, and "T" numbers, such as T.69.

Deutsches Institut für Normung (DIN). In Germany, DIN developed the Werkstoff numbering system for steel in 1959, reserving the 1.4000 to 1.4999 series of numbers for stainless steels. The Werkstoff number 1.4301 is for a stainless steel composition that is similar to type 304. In addition to developing Werkstoff numbers, DIN also developed name designations for each alloy, which consist of chemical symbols and numbers that indicate the approximate percentages of the elements. The name associated with the 1.4301 designation is X5CrNi 18-10. The DIN five-digit numbering system has been generally adopted for the Euronorm numbers. Various forms of this name designation system for stainless steel alloys have been adopted by the national standards organizations of most European countries and the International Standards Organization as well as by Russia and China.

AFNOR (France). In France, there is a dual-designation system for stainless steel consisting of steel names and a numerical system. The names are based on a mixture of abbreviations for chemical symbols and other names. The names start with the letter "Z." The abbreviations include "C" for chromium, "N" for nickel, and "F" for free machining. The name for type 304 stainless steel is Z6CN18-09, and for type 410F it is Z30CF13.

The numerical system uses the AISI number followed by the letter "F" and two digits, such as 304F01.

Swedish Institute for Standards (SIS). In 1947, SIS developed a four-digit numbering system for metals, reserving the 23*xx* series for stainless steels. For example, the number 2332 was established for an alloy having a composition similar to that of type 304.

Japan Industrial Standards (JIS). Japan has generally adopted the AISI numbering system for stainless steels, with some modifications. In the Japanese system, SUS 304 corresponds to type 304 stainless steel. The SUH numbers are used to designate heat-resistant alloys, such as SUH 310. The SCS and SCH numbers are used for stainless steel castings.

ISC (People's Republic of China). China has a dual system of designating stainless steel similar to that used in the DIN and EN systems. It consists of the alphanumerical ISC system and a steel name system based on chemical symbols and numbers indicating percentages.

ISC is the "Iron and Steel Code: Unified Numbering System for Iron, Steel, and Alloys." This ISC system resembles the American UNS system in many respects. It uses a letter followed by five digits; for stainless steels, the initial letter is "S." Some of the entries in the ISC system, such as S31603, S31653, and S41008, are the same as those in the UNS system, and in such cases, the chemical compositions for the alloys are identical in each system. The ISC system for stainless steel groups the alloys into five categories: S1*xxxx* (ferritic), S2*xxxx* (austenitic-ferritic), S3*xxxx* (austenitic), S4*xxxx* (martensitic), and S5*xxxx* (precipitation hardening).

In the steel-naming system, 00Cr17Ni13Mo2N is the designation for the 316LN alloy, which is S31653 in both the ISC and UNS systems.

KS (Republic of Korea). Korea has adopted a numbering system that appears to be consistent with the Japanese system for stainless steel. The prefix letters "STS" are used for corrosion-resisting steels, "STR" for heat-resisting alloys, and "SSC" for stainless steel castings. Therefore, STS 304 corresponds to type 304.

GOST (Russia). Steel names standardized by GOST use chemical symbols and abbreviations of chemical symbols. "Ch" and "K" are used for chromium, "N" for nickel, and "T" for titanium. The name for type 304 is 08Ch18N10.

UNE (Spain). In Spain, there are three numbering systems for stainless steel. The steel names use chemical symbols and numbers, such as X6CrNi 19-10, which also corresponds to the "F" number F.3504.

AISI numbers are also used with the prefix "E," such as E304.

UNI (Italy). Steel names standardized by UNI use chemical symbols and numbers. The name for type 304 is X8CrNi 19-10.

Unified Numbering System (UNS) for Metals and Alloys. The UNS system of numbering metals and alloys is widely used for the designation and cataloging of stainless steels in America and abroad.

The UNS for metals and alloys was developed in 1970 by ASTM and SAE, with the cooperation of the Aluminum Association, AISI, the Copper Development Association, the Steel Founders' Society of America, representatives from the metals industry, and the U.S. government. The goal was to develop a numbering system for metals and alloys that could be used throughout the metals industry. Furthermore, it was to be a system with sufficient capacity to accommodate the possibility of the development of hundreds of new alloys without the need to add suffix letters, as in the case of the AISI numbers for carbon, alloy, and stainless steels. The existing numbers in the aluminum, copper, and steel industries were to become a part of the new system. For stainless steels, the system consists of the letter "S" followed by five digits. For type 304 stainless steel, the UNS number is S30400, and for type 304L, the UNS equivalent is S30403. The system also permits the numbering of stainless steel alloys for which there is no AISI designation, such as S35500 for stainless alloy AM 355. For Ferralium 255, the UNS designation is S32550.

The letter "N" is also used to designate certain stainless alloys that contain less than 50% iron and high nickel contents. The UNS designation for alloy 20Cb-3 is N08020.

For steel casting alloys in the UNS system, a J*xxxxx* series of numbers was established. This "J" series is based on the total alloy content. For stainless steel casting alloys, the numbers fall in a J9*xxxx* series, because the alloys have alloy contents greater than 9%. The ACI alloy CA-15 has the UNS designation J91150. The second, third, and fourth digits of J91150 indicate that the alloy has a minimum alloy content of 11.5%, which happens to be the minimum chromium content for this alloy. (See Chapter 15, "The Unified Numbering System (UNS) for Metals and Alloys.")

International Organization for Standardization (ISO). In approximately 1960, the International Organization for Standardization established ISO Committee TC17 on Steel and Subcommittee SC4 on Heat Treated and Alloy Steel, which developed ISO 683/XIII covering stainless steel chemical compositions. Type 304 stainless steel was designated as type 11 in that document. In 1987, the type numbers for stainless steels were changed to the identical steel names used by DIN. The former type 11 stainless steel is now designated by ISO as X5CrNi 18-10. An exception to this rule is that stainless steels for surgical implants are designated by letters of the alphabet, because these alloys are not included in the DIN system.

In 1980, there was a movement within ISO to develop an international numbering system for metals. The proposal was to use the basic format of the UNS system with an initial letter and five digits. In some cases, the initial letters were the same as those in the UNS system, such as "A" for aluminum and "C" for copper and copper alloys. The letter "S," however, was used to designate steel alloys rather than stainless steels, as in the UNS system. It was obvious to the developers of the UNS system that two systems that looked alike but in fact were different would lead to a great deal of confusion. The ISO numbering system never got beyond the planning stage.

Europäischen Normen (EN). Beginning in 1988, a new series of mandatory European standards (Euronorms) were created to replace national standards, such as British standards, DIN, Swedish standards, and French standards, throughout 18 countries of Western Europe. In 1995, three parts of EN 10088 were published. Part 1 listed the chemical compositions of 83 stainless steels. In 1995, EN 10027-1, "Part 1, Designation Systems for Steels: Steel Names," and EN 10027-2, "Part 2, Designation Systems for Steels: Numerical System," were published.

The designation system for steel names generally adopted the DIN steel names, such as X5CrNi 18-10 for AISI 304 stainless steel, and EN 1.4301 was adopted from DIN Werkstoff number DIN 1.4301, also for the AISI 304 alloy.

APPENDIX 1

Stainless Steel Bibliography

50th Anniversary—The E.G. Budd Manufacturing Company—1912–1962, E.G. Budd Company, Philadelphia, Pa.

Armstrong, P.A.E., Corrosion-Resistant Alloys—Past, Present and Future—With Suggestions as to Future Trend, *Proceedings,* American Society for Testing Materials, Vol 24, 1924, p 193–207

Arnold, W., Vertically Challenged in Malaysia, *New York Times,* New York, N.Y., March 8, 2003

Barraclough, K.C., Sheffield and the Development of Stainless Steel, *Ironmaking and Steelmaking,* Vol 16 (No. 4), The Institute of Metals, London, 1989, p 253–265

Brearley, A., and Brearley H., *Ingots and Ingot Moulds,* Longmans, Green and Co., 1918

Brearley, H., Cutlery, U.S. Patent 1,197,256, Sept. 5, 1916

Brearley, H., *Harry Brearley—Stainless Pioneer,* British Steel Stainless and The Kelham Island Industrial Museum, Sheffield, England, 1989

Brearley, H., Steel Production, Canadian Patents 164,622, 1915, and 193,550, 1919

Bringas, J., and Lamb, S., *Stainless Steel and Nickel Alloys,* Casti Publishing, Inc., Edmonton, Alberta, Canada, and ASM International, Materials Park, Ohio, 1999

Burlington Zephyr Smashes Scot's Record in Non-Stop Run Denver to Chicago, *Philadelphia Inquirer,* May 27, 1934

Carpenter Technology, 100 Years of Progress, Carpenter Technology, Reading, Pa., 1989

Castro, R., Historical Background of Stainless Steels, *Les Aciers Inoxydables,* Les Éditions de Physique, Les Lulis, France, 1990, p 1–9

Cobb, H.M., *ASTM and SAE-AMS Stainless Steel Specifications,* ASTM International, W. Conshohocken, Pa., 2003

Cobb, H.M., Development of the Unified Numbering System for Metals and Alloys, *Standardization News,* American Society for Testing and Materials, W. Conshohocken, Pa., Feb. 2002

Cobb, H.M., *Stainless Steels,* The Iron and Steel Society, Warrendale, Pa., 1999

Cobb, H.M., *Stainless Steels,* The Association for Iron and Steel Technology, Warrendale, Pa., 2008

Cobb, H.M., *Stainless Steel Specifications,* Specialty Steel Industry of North America (SSINA), Washington, D.C., 2008

Cobb, H.M., The Naming and Numbering of Stainless Steels, *Advanced Materials and Processes,* ASM International, Materials Park, Ohio, 2007

Coel, M., A Silver Streak, *Invention and Technology,* Vol 2 (No. 2), Fall 1986

Columbier, L., and Hochmann, J., *Stainless and Heat Resisting Steels,* Edward Arnold Publishers Ltd., London, 1965, p 3–6

Davis, J.R., *Stainless Steels,* ASM International, Materials Park, Ohio, 1994

Davis, J.R., Ed., Stainless Steels, *An Alloy Digest Sourcebook,* ASM International, Materials Park, Ohio, 2000

Dunlap, D.W., Juke Box in the Sky, *New York Times,* May 26, 2005, p F1

Dupré, J., Chrysler Building, *Skyscrapers: A History of the World's Most Famous and Important Skyscrapers,* Black Dog & Leventhal, New York, N.Y., 1996, p 36–37

Eduard Maurer and His Contribution to the Development of Stainless Steel, *Korrosions-Beständige Stähle*, Deutscher Verlag für Grundstoffindustrie GmbH, Leipzig, 1990, p 12–17

Feild, A.L., Manufacture of Stainless Iron from Ferrochromium, from Scrap, or from Ore, *Metals Progress,* Feb. 1933

Firth-Sterling "S-Less" Stainless Steel, Firth-Sterling Steel Co., McKeesport, Pa., 1923

Gray, R.D., *Alloys and Automobiles—The Life of Elwood Haynes,* Indiana Historical Society, Indianapolis, Ind., 1979

Guillet, L., *Special Steels,* Dunod, Paris, 1905

Hampel, C.A., *Rare Metals Handbook,* 2nd ed., Reinhold Corp., London, 1961

Hatfield, W.H., Heat Treatment of Aircraft Steels, *British Journal of Automobile Engineers,* Vol 7, 1917

Haynes, E., Stellite and Stainless Steel, *Proceedings,* Vol 35, Engineers' Society of Western Pennsylvania, Pittsburgh, Pa., 1920, p 467–482

Horsely, C.B., *The Chrysler Building,* The Midtown Book, New York, N.Y., 1998

Houska, C., Stone, P.G., and Cochrane, D.J., "Timeless Stainless Architecture," Nickel Development Institute, Toronto, 2001

Hua, M., Tither, G., et al., Dual Stabilized Ferritic Stainless Steel for Demanding Applications Such as Automotive Exhaust Systems, *Iron and Steel Magazine,* Warrendale, Pa., April 1997, p 41–44

Hunt, J.M., Stainless Steel in Automobiles, *Advanced Materials and Processes,* ASM International, Materials Park, Ohio, May 1996, p 39–40

Innovation Stainless Steel, Florence, Italy, 1993, p 1.103–1.108

Jonas, E.A., One Hundred Years—Committee A-1 on Steel, Stainless Steel and Related Alloys, *Standardization News,* American Society for Testing and Materials, W. Conshohocken, Pa., Feb. 1998, p 21–23

Kerr, J.W., *Illustrated Treasury of Budd Railway Passenger Cars,* Delta Publications, Alburg, Vt., 1981

Krainer, H., 50 Jahre Nictrostender Stahl, *Techn. Mitt. Krupp · Werksberichte,* Vol 20 (No. 4), 1962, p 165–179

Krivitsky, W.A., The Linde Argon-Oxygen Process for Stainless Steel: A Case Study of Major Innovation in a Basic Industry, *Metallurgical Transactions,* Vol 4, June 1973, p 1439–1447

Lamb, S., and Bringas, J.E., *Practical Handbook of Stainless Steel and Nickel Alloys,* Casti Publishing, Inc., Edmonton, 1999

Langehenke, H., Dusseldorf, Steel Institute VDEh, private correspondence, Jan. 28, 2003

Long Life Ambition—Stainless Steel Reinforcing Used for Guildhall Yard, *NiDI,* Vol 1 (No. 1), 1996

Lula, R.A., *Stainless Steel,* American Society for Metals, Metals Park, Ohio, 1986

MacQuigg, C.E., Some Engineering Applications of High-Chromium-Iron Alloys, *Proceedings,* American Society for Testing Materials, Philadelphia, Pa., 1924, p 373–382

Magee, J H., and Schnell, R.E., Stainless Steel Rebar, *Advanced Materials and Processes,* ASM International, Materials Park, Ohio, Oct. 2002, p 43–45

Mahlberg, G., DeLorean: Stainless Style, *DeLorean World,* Vol 15 (No. 3), 1995

Marble, W.H., Stainless Steel—Its Treatment, Properties and Applications, *Transactions of the American Society for Steel Treating,* Vol 1, Cleveland, Ohio, 1920, p 170–179

McIntyre, J., A Lifetime's Passion for Stainless Steel (an interview with Harold M. Cobb), *Stainless Steel World,* Nov. 2009, p 2–3

Megerian, C., and Mosgrove, C., The Chrysler Building: An Engineering Reform and an Architectural Revolution, *Building Big: Skyscrapers,* PBS, WGBH Television, Boston, Oct. 17, 2000

Miettinen, E., Architectural Design with Stainless Steel in Europe, *Symposium on Stainless Steel in Architecture,* Berlin, 2000

Monypenny, J.H.G., *Stainless Iron and Steel,* Chapman & Hall, Ltd., London, 1926

"New Bridge Using Stainless Steel Rebar to Last 120 Years in Corrosive Marine and Earthquake Environment," Talley Metals, 2002

Nisbett, E.G., A Partnership That Saves Lives, *Standardization News,* American Society for Testing and Materials, W. Conshohocken, Pa., Feb. 1998, p 24–27

Olsson, J., and Liljas, M., 60 Years of Duplex Stainless Steel Applications, *Corrosion 94,* NACE International, Houston, Texas, 1994, p 395-1 to 395-12

Parmiter, O.K., in *Transactions of the American Society for Steel Treating,* Vol VI, 1924, p 315–340

Patil, B.V., Chan, A.H., and Choulet, R.J., Refining of Stainless Steel, *The Making, Shaping and Treating of Steel,* The AISE Steel Foundation, Pittsburgh, Pa., 1998, p 715–718

Pickering, F.B., *The Metallurgical Evolution of Stainless Steels,* American Society for Metals, Metals Park, Ohio, and The Metals Society, London, 1979

Pierpont, C.R., The Silver Spire, *The New Yorker,* New York, N.Y., Nov. 18, 2002, p 74–81

Plummer, C.E., Classification of Trade Names, *The Book of Stainless Steels,* E. Thum, Ed., American Society for Metals, Cleveland, Ohio, 1935, p 713–737

Pollock, W.I., Some Historical Notes, *Practical Handbook of Stainless Steels and Nickel Alloys,* ASM International, Materials Park, Ohio, and Casti Publishing, Edmonton, Alberta, Canada, 1999, p 1–29

Ragsdale, E.J.W., *Testing of Railway Passenger Cars and the Materials of Construction,* E.G. Budd Manufacturing Company, Philadelphia, Pa., 1941

Ragsdale, E.J.W., *Testing of Railway Passenger Cars and the Materials of Construction,* Edward G. Budd Manufacturing Company, Philadelphia, Pa., 1950

Ragsdale, E.J.W., and Dean, A.G., Rail Car Body (stainless steel), U.S. Patent 2,129,235, Sept. 6, 1938

Report on Inspection of Corrosion-Resistant Steels in Architectural and Structural Applications, *Proceedings,* American Society for Testing Materials, Philadelphia, Pa., 1962, p 188–193

Reuter, M., The Lost Promise of the American Railroad, *The Wilson Quarterly,* Washington, D.C., Winter 1994

Reynolds, J.B., *The Chrysler Building,* New York, N.Y., 1930

Schoefer, E.A., Evolution of the ACI Alloys, *Steel Foundry Facts,* 1981, p 13–18

Scott, S., The Savoy Group, London, private correspondence, Nov. 4, 2002

Snelgrove, P., "Stainless Steel Automotive and Transport Developments," World Stainless Library, 2001

Snyder, C.C., Finishing Stainless Steels, *Metal Progress,* Cleveland, Ohio, Feb. 1931, p 77–80

Stainless Steels, *The Making, Shaping and Treating of Steel,* 8th ed., United States Steel Corp., 1964, p 1111–1130

Strauss, B., Non-Rusting Chromium-Nickel Steels, *Proceedings* (Atlantic City, N.J.), American Society for Testing Materials, Philadelphia, Pa., 1924, p 208–216

Strauss, J., Introduction—With Tabulation of Manufacturers' Data on Composition and Properties of the Alloys, *Proceedings* (Atlantic City, N.J.), American Society for Testing Materials, Philadelphia, Pa., 1924, p 189–192

Stravitz, D., Ed., *The Chrysler Building: Creating a New York Icon a Day at a Time,* The Princeton Architectural Press, N.Y., 2002

Streicher, M.A., Stainless Steels: Past, Present and Future, *Stainless Steel '77,* Climax Molybdenum Company and Amax Nickel, 1977, p 1–2

Stubbles, J.R., *Original Steelmakers,* Iron and Steel Society, Warrendale, Pa., 1984

The Chrysler Building, Chrysler Tower Corporation, New York, N.Y., 1930

The Making, Shaping and Treating of Steel, 10th ed., Association for Iron and Steel Engineers, Pittsburgh, Pa., 1985

The New Stainless Steel, *Scientific American,* March 31, 1917, p 329

The Pioneer Zephyr, American Society of Mechanical Engineers, New York, N.Y., 1980

Thum, E.E., Ed., *The Book of Stainless Steels,* 1st ed., The American Society for Steel Treating, 1933
Thum, E.E., Ed., *The Book of Stainless Steels,* 2nd ed., American Society for Metals, Cleveland, Ohio, 1935
Tverberg, J.C., Stainless Steel in the Brewery, *MBAA Technical Quarterly,* Vol 38 (No. 2), 2001
Tyson, S.E., The U.S. Experience in ISO Technical Committee TC/17 on Steel, *Standardization News,* American Society for Testing and Materials, W. Conshohocken, Pa., Feb. 1998, p 40–43
Van Alen, W., Architectural Uses, *The Book of Stainless Steels,* American Society for Metals, Cleveland, Ohio, 1935
Watter, M., and Lincoln, R.A., *Strength of Stainless Steel Structural Members as a Function of Design,* Allegheny Ludlum Steel Corporation, Pittsburgh, Pa., 1950
Zapffe, C.A., *Stainless Steels,* American Society for Metals, Cleveland, Ohio, 1949
Zapffe, C.A., *The Fascinating History of Stainless Steel—The Miracle Metal,* Republic Steel Corporation, Cleveland, Ohio, 1960
Zapffe, C.A., *Trans. ASM,* Vol 34, 1945, p 71–107
Zephyr Makes World Record Run, 1017 Miles at Average of 78 an Hour, *New York Times,* May 26, 1934

APPENDIX 2

A Stainless Steel Timeline

1751 **Nickel discovered.** Nickel is discovered in Sweden by Axel Cronsted in kupfernickel (nicolite), a copper-colored nickel arsenide mineral.

1774 **Manganese discovered.** Manganese is discovered by Johann Gann in Sweden and recognized as an element by the Swedish chemist Karl Wilhelm Scheele.

1778 **Molybdenum discovered.** Swedish scientist Karl Wilhelm Scheele conducts research on an ore now known as molybdenite. He concludes that it does not contain lead, as suspected at the time, and reports that the mineral contains a new element, which he calls molybdenum.

1781 **Tungsten discovered.** Swedish chemist Karl Wilhelm Scheele first identifies tungsten in a calcium tungstate mineral, which later becomes known as scheelite. The word *tungsten* is an adaptation of the Swedish *tung sten*, meaning heavy stone.

Tungsten also became known in Europe as wolfram, which is a pejorative term based on the German *wolf* plus *ram*, meaning dirt. It is explained that tin miners gave the material this term because it was considered a worthless material. Therefore, "W" became the chemical symbol for wolfram. The reason for the more common use of the name tungsten is explained in the 1844 entry on columbium, another contentious name.

1782 **Molybdenum metal prepared.** Molybdenum metal is prepared in an impure form by the Swedish scientist Peter Jacob Hjelm. (The Climax Mine in Climax, Colorado, has been the world's largest molybdenum mine.)

1790 **Titanium discovered.** Titanium is first discovered in England by Wilhelm Gregor, although it did not receive a name until Martin Heinrich Klaproth found it in Hungary in 1795 and named it after the mythological first sons of the earth, the Titans. Gregor discovered titanium in iron-bearing sands of the type known today as ilmenite.

1797 **Chromium discovered.** Chromium is discovered by German scientist M.H. Klaproth and French analyst Louis Nicolas Vauquelin, who isolates the metal in 1798. Its name is derived from the Greek word *chromos*, meaning color, because of the various colors of its compounds. Virtually all ores of chromium are made up of the mineral chromite.

1798 **First chromium metal.** The world's first piece of chromium metal is presented to the French Academy by Louis Nicolas Vauquelin. His presentation notes "the metal resists acids, surprisingly well."

1801 **Columbium discovered.** In approximately 1734, the first governor of Connecticut, John Winthrop the younger, discovered a new mineral that he called columbite. (Columbia is a synonym for America.) A sample was sent to the British Museum. The sample was finally examined in 1801 by the British chemist and manufacturer Charles Hatchett, who discovered in it a new element that he called columbium, because Winthrop had named the mineral columbite. Columbium is rediscovered and renamed in 1844. (See also 1844, columbium rediscovered.)

1802 **Tantalum discovered.** Tantalum, named for Tantalus, is discovered by Anders Ekeberg and isolated by Jöns Berzelius. The word *tantalize* was named after Tantalus, and it is believed that tantalum was so named because of the tantalizing problems posed by the inertness of the element and its compounds.

1803 **Vanadium discovered.** Andres Manuel del Rio, a professor of mineralogy at the School of Mines in Mexico City, discovered an element in lead ore that he first called brown lead. He later changed the name to erythronium, because the salts of the material became red when heated.

In 1831, Nils Gabriel Sefström, a Swedish chemist, isolated what he thought was an unknown metal in the iron ores of Taberg, Sweden. He called the material vanadium in honor of Vanadis, the goddess of beauty and fertility in Scandinavian mythology.

Friedrich Wôhler, a German chemist, demonstrated that erythronium and vanadium were the same element. In 1896, the Firminy Steel Works in France first used vanadium in steel. They made three plates of armor that had remarkable properties. However, there was no adequate supply of vanadium in sight until 1905, when Riza Patron, a prospector, discovered the first really large deposit of vanadium ore in the Peruvian Andes at an altitude of 16,000 feet. This source furnished most of the world's requirements for many years.

1811 **Boron metal prepared.** J. Gay-Lussac and L. Thenard prepare impure boron in France by the reduction of boron trioxide.

1811 **Silicon prepared.** J. Gay-Lussac and L. Thenard prepare what was probably impure silicon by passing the trichloride over heated potassium.

1811 **Krupp founded.** Friedrich Krupp founds the first works for cast steel in Germany.

1821 **Berthier.** Pierre Berthier, a Frenchman, writes a paper, "On the Alloys of Chromium with Iron and with Steel," that appears in *Annales de Chemie et de Physique*. Berthier worked with high-carbon chromium and iron and noticed that the higher the chromium content of his steels, the greater the resistance to corrosion. Berthier produced a steel containing only 1.0 to 1.5% chromium and made a knife and razor blades of fine quality from it. However, Berthier worked with steels of high carbon content and with chromium contents either too high or too low to have the unique properties of stainless steel.

1822 **Stodart and Faraday.** Stodart and Faraday find improved atmospheric corrosion resistance with 3% chromium steel. However, dilute sulfuric acid attack was worse than with "regular steel."

1824 **Silicon purified.** Jon Berzelius purifies silicon in Sweden by the same general method as Gay-Lussac in 1811.

1825 **Aluminum discovered.** Danish chemist Hans Christian Oersted (1777–1851) produces an aluminum amalgam in 1825 but is unable to purify the metal. Friedrich Wöhler (1800–1882), a German chemist, isolated the metal in an impure form. It was named aluminum because it was found in alum. Two years later, the name was changed to aluminium to make it consistent with other metallic elements. In 1925, the American Chemical Society changed the name to aluminum, which is the usage in North America. Elsewhere, it is aluminium.

Aluminum was much too expensive for general use. It was considered a precious metal until 1885, when Charles Martin Hall in America invented a process of passing an electric current through a molten bath of the ore to produce pure molten aluminum, which settles at the bottom and is siphoned out. Paul L.T. Heroult, a French chemist, invented the same process in the following year.

1838 **Mallet.** R. Mallet presents a paper entitled "The Action of Water on Iron" before the British Association for the Advancement of Science. He shows, as had Stodart and Faraday in 1822, that chromium-iron alloys or chromium steels are very resistant to oxidizing agents, more so than other alloys tested. However, he comes to the strange conclusion that the chromium probably finally leaches out, leaving the metal more corrodible than ever.

1842 **Thomas Firth & Sons.** Thomas Firth & Sons Steel, Ltd. is established at Sheffield, England. They will produce the first commercial heat of chromium stainless steel in 1913.

1844 **Columbium rediscovered.** Heinrich Rose rediscovers what Hatchett had called columbium in 1801, giving it the new name niobium, for the goddess Niobe, whose father was the mythical king Tantalus. Many scientists thought niobium (columbium) and tantalum were identical elements until Rose, in 1844, and Marignac, in 1866, showed that niobic and tantalic acids were two different acids.

Over the next 100 years, the name columbium was generally used in the United States, and niobium was generally used in Europe. In 1951, after 100 years of controversy, the International Union of Pure and Applied Chemistry (IUPAC) officially adopted niobium as the name for element 41. This was a compromise wherein the IUPAC accepted tungsten instead of wolfram, in deference to North American usage, and niobium instead of columbium, in deference to European usage. Not everyone agreed, however, and while many leading chemical societies and government agencies in the United States use the official IUPAC name, many leading metallurgists, metal societies, and most leading American producers still refer to the metal as columbium. (The principal source of the metal is the mineral columbite.)

1854 **Chromium resists acids.** W. Bunsen, a German, notes that unalloyed chromium metal resists the strongest acids, including aqua regia.

1862 **Sandvik Steel.** The Sandvik Steel Company is founded in Sandviken, Sweden.

1863 **Metallography.** Dr. Henry Clifton Sorby (1826–1908) in England pioneers the field of microscopic metallurgy and is the first to prepare pictures of metals at high magnifications that show their crystalline structures. He named seven crystal structures that he observed in iron and steel. (The development of metallography was most important to stainless steel technology.)

1866 **Columbium metal prepared.** Columbium metal is said to have been prepared first by Christian Wilhelm Blomstrand, who reduced columbium chloride with hydrogen.

1869 **British Iron and Steel Institute.** The British Iron and Steel Institute is formed, with 292 members.

1869 **Chromium alloy steel.** J. Baur establishes the Chrome Steel Works in Brooklyn, New York. He produced what was apparently the first steel that contained small amounts of chromium. Baur advertised that he supplied chromium steel for the famous Eads Bridge, which spanned the Mississippi River at St. Louis. The Eads Bridge, which is still standing, is the oldest steel girder bridge in America. Baur was awarded U.S. Patent 49,495 for "The Manufacture of Steel."

1871 **Thyssen & Company.** August Thyssen founds Thyssen and Company in Styrum (today Muylhelm Styrum), Germany.

1871 **Woods and Clark.** John T. Woods and John Clark, English scientists, recognize the commercial value of corrosion-resistant chromium alloys and obtain a British patent for "Weather-Resistant Alloys." They stated that, for all ordinary purposes, use approximately 5% chromium with a small addition of tungsten with medium carbon and, for purposes requiring maximum resistance to corrosion, use a 30% chromium steel with approximately 1.5% tungsten. They also reported that the higher-chromium-content alloy is particularly useful for cutlery, drawing instruments, coinage metal, and metal mirrors.

In a paper presented at the 27th Annual Meeting of the American Society for Testing Materials in 1924, P.A.E. Armstrong of the Ludlum Steel Company said the following with regard to the work of Woods and Clark:

"These two inventors who, it is believed, were the first to give practical information on the limits of composition of rustless and stainless iron and steel, styled themselves in the specifications a chemist and engineer. Our knowledge of present-day rustless iron does not go very much further than the disclosure of these two men."

1875 **Ludlum Steel.** In 1875, the Pompton Steel & Iron Company is formed at Pompton Lakes, New Jersey, with James Ludlum as president. Upon James Ludlum's death in 1892, his son, William, succeeded him as general manager of the company, which was renamed Ludlum Steel and Spring Company. When Ludlum's dam broke in 1903, the plant, which had become completely obsolete, was severely damaged. Erastus Corning came to Ludlum's rescue. Mr. Corning decided to build a new plant at Watervliet, New York. In 1907, the company was moved to Watervliet, and in 1910, the company was renamed Ludlum Steel Company. Ludlum produced one of the first commercial heats of chromium stainless steel in America in 1918. Ludlum and Allegheny Steel merged on August 16, 1938, to form Allegheny Ludlum Steel Corporation, which created the largest steel company specializing in stainless, electrical, tool, and other alloy steels and carbides.

1877 **Chromium steels.** J.B. Boussingalt and Almé Brustlein, working at Aciéries Holtzer in Unieux, France, develop chromium steels for the first time in Europe. They recognize that the mechanical properties are dependent upon the chromium and carbon contents.

1878 **Chromiferous spiegeleisen.** The Terre Noire Company in France produces a *chromiferous spiegeleisen*, an iron alloy with a brilliant crystalline fracture (hence the German name for brilliant iron) with 25% chromium and 13% manganese.

1886 **Chromium steels.** It becomes common knowledge among analytical chemists that chromium-containing steels exhibit greater corrosion resistance in many media.

1886 **Carpenter Steel.** The Carpenter Steel Company is established at Reading, Pennsylvania.

1892 **Hadfield.** Sir Robert A. Hadfield studies 1 to 9% chromium alloys with 1 to 2% carbon in 50% sulfuric acid solutions and concludes that chromium is deleterious. Hadfield is probably best known as the inventor of Hadfield's manganese steel, a steel with 11% manganese that is fully austenitic but transforms to martensite when cold worked and becomes highly abrasion resistant. (Note: R.A. Hadfield is not to be confused with Dr. William H. Hatfield, the English metallurgist who became the inventor of the 18-8 stainless steels in 1924.)

1895 **Low-carbon ferrochromium.** In Germany, Hans Goldschmidt patents a method for preparing carbon-free chromium. Low-carbon ferrochromium and chromium metal are produced.

1898 **High-carbon alloys.** In France, A. Corot and E. Goutal discover the damaging effect of high carbon content on the corrosion resistance of iron-chromium alloys.

1898 **ASTM.** The American Society for Testing Materials (ASTM) is established at Philadelphia, Pennsylvania. Committee A-1 on Steel became the first of more than 150 technical committees. (The name was changed in the 1950s to The American Society for Testing and Materials.)

1899 **Armco.** The American Rolling Mill Company is established at Middletown, Ohio. The name was later changed to the Armco Company and then to the A-K Steel Company in 1994.

1900 **Crucible Steel Co.** The Crucible Steel Company of America is incorporated, with its headquarters in Pittsburgh, Pennsylvania.

1900 **Holtzer Company.** Holtzer and Company of Unieux, France, exhibit stainless steel at the Paris Exposition, which is described in the 1924 *Proceedings of the American Society for Testing Materials* as follows:

"The Holtzer Exhibit was extremely interesting from the point of view of the range of analyses and physical properties in the annealed, hardened and drawn condition. These analyses are right in the range of present-day rustless steel and cover the so-called stainless steels. It is perhaps fair to say that their non-rusting and acid-resisting properties must have been observed, although perhaps not chronicled."

1901 **Allegheny Steel and Iron.** The Allegheny Steel and Iron Company is incorporated at Pittsburgh, Pennsylvania, and a plant is built at Brackenridge, Pennsylvania.

Allegheny Steel and Iron Company and Ludlum Steel merged on August 16, 1938, to form Allegheny Ludlum Steel Corporation, which created the largest steel company specializing in stainless, electrical, tool, and other alloy steels and carbides.

1901 **BSI.** The British Standards Institute (BSI) is established in London.

1902 **US Steel Co.** The United States Steel Company (USS) is created by a merger of J.P. Morgan's Federal Steel Company with Andrew Carnegie's Carnegie Company at Pittsburgh, Pennsylvania.

1902 **Jones & Laughlin.** Jones & Laughlin Company is incorporated at Pittsburgh, Pennsylvania.

1902 **INCO.** The International Nickel Company, Ltd. (INCO) is created in Camden, New Jersey, as a joint venture between Canadian Copper, Orford Copper, and American Nickel Works.

1902 **Nickel refinery.** A nickel refinery is completed at Clydach, Wales, by the Mond Nickel Company.

1903 **Rustless steel.** British Patent 23,681, issued to La Societé Anonyme de la Neo-Metallurgie, is for rustless medium-carbon steel. The inventors used iron, nickel, and chromium, with iron from 16 to 38%, nickel from 5 to 60%, and chromium from 24 to 57%. The following is quoted from the patent:

"But the most important modification to which these elements, nickel and chromium, give rise consists in the important property (which forms the subject of researches at the present time) which they communicate to the said products, steels and irons.

"This quality adapts them for a number of purposes and uses for which irons containing carbon and manganese in greater or lesser proportions are only very imperfectly adapted by reason of their liability to corrosion which is attributed to a great extent to the manganese which they contain."

As is known today, these alloys are very rust-resisting and representative of very good rustless iron and steel.

1904 **Guillet.** Léon B. Guillet, professor of metallurgy and metal processing at the Conservatoire des Arts et Métiers, publishes a series of research on iron-chromium alloys having carbon contents acceptably low for modern stainless steel analyses. He made his steels with carbon-free Goldschmidt chromium (oxide ore reduced to chromium with powdered aluminum). He investigates iron alloys with 12 to 26% chromium.

Guillet's article appears in 1904 in *Le Genie Civil* and discusses, from a metallographic point of view, a large number of analyses similar to those of the Holtzer Exhibit in 1900. He notes that the alloy of a composition similar to that of present-day rustless steel could not be etched in nitric and picric acids, and he was forced to use other acids, such as hydrochloric, to etch them. This is the first known reference to the use of metallography in connection with stainless steel.

In 1906, he also publishes work including nickel at present-day percentages.

1905 **First book on stainless steel published.** *Stainless Steel* is authored by Léon Guillet in Dunod, Paris. It has 132 pages.

1906 **Guillet studies iron-chromium-nickel alloys.** Léon Guillet publishes a detailed study of iron-chromium-nickel alloys that establishes the basic metallurgical characteristics of the chromium-nickel stainless steels.

1907 **American Steel Founders' Society organized**. The American Steel Founders' Society of America (ASFS) is organized in New York City, with O.P. Letchworth as its first president.

1907 **Brown Firth Laboratories.** Thomas Firth & Sons and John Brown & Company, two neighboring steelmakers in Sheffield, set up a joint development department. Harry Brearley is asked to become the first head of the Brown Firth Research Laboratories.

1908 **AISI is organized.** The American Iron and Steel Institute (AISI) is organized in New York City on March 31, 1908.

1909 **Giesen and Portevin.** Between 1909 and 1912, W. Giesen and Albert Portevin, in France, publish work on the three types of stainless steel then known, which were roughly equivalent to the modern austenitic, martensitic, and ferritic stainless steels.

1910 **Iron rusts.** According to *The Corrosion and Preservation of Steel*, a journal published in England by Cushman and Gardner, "The tendency to rust is inherent in the element known as iron and will, in all probability, never be entirely overcome."

1910 **Ludlum electric furnace.** Ludlum Steel of Watervliet, New York, is the first to use the electric furnace in melting alloy steel.

1910 **Borchers and Monnartz.** W. Borchers and Philip Monnartz, in Germany, obtain German Patent 246,015 on a stainless steel.

1911 **Monnartz study.** Philip Monnartz, in Germany, publishes a classic work, "The Study of Iron-Chromium Alloys with Special Consideration of their Resistance to Acids," which appears in *Metallurgie.* Monnartz is the first to explain that stainless steel requires at least 12% chromium and a controlled amount of carbon, but it will be many years before the chromium oxide layer theory is put forth.

1911 **Dantsizen.** Christian Dantsizen of the General Electric Laboratories, Schenectady, New York, develops a low-carbon, iron-chromium ferritic alloy having 14 to 15% chromium and 0.07 to 0.15% carbon for use as leading-in wire for electric bulbs.

1911 **Haynes patent.** Elwood Haynes, founder of the Haynes Stellite Company, Kokomo, Ind., experiments with five iron-chromium alloys that he finds to be resistant to corrosion. He applies for a patent, which at first is denied because the U.S. Patent Office already had patents for chromium steels. Also, because the patent claims were similar to those of Harry Brearley, a patent is not granted to Haynes until April 1, 1919.

1912 **Maurer and Strauss.** In 1909, Eduard Maurer joined the research laboratory at the Friederich A. Krupp Works at Essen, Germany, as their first metallurgist. In 1912, while searching for alloys suitable for use in pyrometer tubes, Maurer discovers that some iron-chromium alloys with approximately 20% chromium and 8% nickel, which Dr. Benno Strauss had made, are impervious to attack after months of exposure to acid fumes in his laboratory.

Patents in 1912, and a paper presented by Strauss before a meeting of chemists in Bonn in 1914, defined these alloys as comprising the following two types:

- 7.0 to 25% chromium, 1% maximum carbon, and 0.5 to 20% nickel (German Patent 304,126)
- 15 to 40% chromium, 1.0% maximum carbon, and 4 to 20% nickel (German Patent 304,159)

The first type had to do with a martensitic grade, V1M (with "M" for martensite), containing 14% chromium and 2% nickel, similar to England's Two Score alloy (20% chromium and 2% nickel) and our modern types 414 (12% chromium and 2% nickel) and 431 (16% chromium and 2% nickel), and was recommended for general corrosion resistance.

The second type was an austenitic grade V2A (with "A" for austenite), containing 20% chromium and 7% nickel, which was similar to the 18% chromium and 8% nickel, or 18-8, of today, and was recommended for its exceptional corrosion resistance, particularly in nitric acid.

Articles of both types of steel were displayed at the Mälmo Exhibition in Sweden in 1914.

Krupp supplied the V2A steel to Badishe Analin und Sodafabric for the production of ammonia. (See also 1914, Malmo Exposition.)

1912 **Thompson-Houston patent.** The British Thompson-Houston Company is granted British Patent 15,342 for stainless steel lead-in wires for glass bulbs for electric lamps, claiming that the oxide formed on the surface of the high-chromium wire was very thin and that it welded nicely with the glass. (Here was recognition of the heat-resisting properties of high-chromium, low-carbon irons.)

1912 **Schilling patent.** Schilling receives U.S. Patent 1,026,461 for a chromium steel with approximately 9% chromium, 3% molybdenum, and approximately 1% carbon. It was to be employed as a safe vault steel, and it was claimed that the steel could not be melted by an oxyacetylene flame.

1913 **Krupp patents.** Patents covering the manufacture of two iron-chromium-nickel alloys are taken out in England and other countries by Clement Pasel, a firm associated with Krupp. The patents are held by Krupp. The British Patent 13,414 applies to general corrosion resistance but not to acids. Patent 13,415 applies to chromium-nickel alloys resistant to the attack of acids, with nitric acid being especially mentioned. Krupp manufactured steel falling within the first patent as V.1.M., where the "M" denotes martensitic, and a steel within the second patent as V.2.A., where the "A" denotes austenitic. The V.2.A alloy was a 20% chromium and 7% nickel alloy that was used for many years before the 18-8 alloy became the most popular.

The names of Strauss and Maurer are not mentioned in the patents.

1913 **Strauss applies for U.S. patent.** An application is filed on June 25, 1913, by Dr. Benno Strauss of the German Empire for an American patent for "Articles which Require High Resistance against Corrosion." U.S. Patent 1,316,817 was assigned to the Chemical Corporation, Inc., a corporation of Delaware, and granted on September 23, 1919.

Strauss stated, "This invention relates to articles such as gun barrels, turbine blades and the like, which contain 6% to 25% chromium and from 90% to ½% nickel and not over 1% carbon."

The patent covers Krupp alloy V.2.A., the 20% chromium and 7% nickel alloy that was studied in the Krupp investigation and was widely used for many years, until the development of the 18-8 alloy.

1913 **Brearley's discovery.** Harry Brearley, who was the director of the Brown Firth Research Laboratories in Sheffield, England, is generally credited above all others as the initiator of the industrial era of stainless steel, at least as far as the hardenable chromium stainless steels are concerned. In 1913, while he was investigating the development of new alloys for rifle barrels, he noted that the customary etching reagents for steel did not etch steels having high chromium contents. He concluded that a steel with more than 10% chromium may be of advantage in his particular application.

The first commercial heat of chromium stainless steel is cast on August 20, 1913, in Sheffield, England, by Harry Brearley. Its composition is 12.86% chromium and 0.24% carbon, a martensitic steel that was the forerunner of type 420 stainless steel. It seems that the alloy did not show the desired improvement with regard to gun barrels; however, material from the same cast was eventually made into cutlery blades. This was the first commercial cast of what was subsequently known as stainless steel.

In a report dated October 2, 1913, Brearley wrote, ". . . these materials would appear especially suited for the manufacture of spindles for gas and water meters, pistons and plungers in pumps, ventilator and valves in gas engines and, perhaps certain forms of cutlery."

Brearley obtained U.S. and Canadian patents in 1915. (See also entries for 1915 and 1916.)

1914 **Dantsizen.** Christian Dantsizen of General Electric Research Laboratories, Schenectady, N.Y., extends the use of his ferritic 14 to 16% chromium and 0.07 to 0.15% carbon stainless steel to steam turbine blades.

1914 **Becket.** Frederick M. Becket of Electro Metallurgical Company, Niagara Falls, New York, looks for cheap alloys for heat and scale resistance in a tunnel oven. Knowing through ferrochromium manufacture that the alloys are remarkably nonscaling, he determines that a minimum of 20% chromium is needed, with relatively low carbon. Becket produced an oxidation-resistant "chrome iron" containing 25 to 27% chromium that, although having a high carbon content, was a ferritic stainless steel, because the chromium was too high to allow significant hardening by heat treatment. Becket also invented low-carbon ferrochromium, which he produced by reducing the chromium oxide with silicon rather than aluminum.

1914 **Silicon-chrome steels.** P.A.E. Armstrong of the Ludlum Steel

Company, Watervliet, New York, discovers the silicon-chrome steels, which are principally used for gasoline engine exhaust valves, by an accidental contamination of a small electric furnace by some silicon reduced from an asbestos cover on the electrode. This alloy was trademarked Silchrome. The American Stainless Steel Company sued Ludlum Steel for infringement of the Brearley and Haynes patents held by them. American Stainless won the suit but the decision was highly questionable.

1914 **Malmo Exposition.** Goods made from the first stainless steels patented by Krupp are displayed in the spring at the Baltic Exposition at Malmo, Sweden. The following excerpts are from the Krupp brochure handed out at the exposition:

"Up until now, high-content nickel steels were regarded as the least rusting steel alloys. However, the new patented rust-free steels—also highly alloyed steel—by far surpass the nickel steels regarding rust resistance as well as resistance to all kinds of corrosion, and additionally have outstanding qualities regarding strength. Two types are produced with essentially different physical qualities and different structures. Brand name V1M, with high stretching limits, should be chosen for machine parts with high mechanical demands. For machine parts and devices which are exposed to chemical influences and have to be very corrosion resistant, brand name V2A is recommended which can also be considered to be absolutely rust free in damp air."

The brochure goes on to note that uses of V1M are as follows: "rifle barrels, turbine blades, parts for pumps, valves of all kinds, parts for ignitions devices, piston rods and knife blades." V2A steel is said to be "especially suited to various uses by the chemical industry where resistance to saltpeter salts or ammonia, together with water vapors, is of prime importance."

Finally, it is stated that "these steels are not sufficiently resistant to the attack by sulphuric or hydrochloric acids."

1914 **Cutlery steel produced.** Thomas Firth & Sons, Ltd., Sheffield, produces 50 tons of cutlery stainless steel.

1914 **Stainless table knives first made.** Table knife blades are forged from the first chromium steel cast by Thomas Firth & Sons of Sheffield. The blades are made of Harry Brearley's 12% chromium steel and are stamped with the following information:

"STAINLESS KNIFE; George Ibberson & Co. Sheffield Eng; Firth-Brearley"

1914 **Microexamination.** In June 1914, Dr. Benno Strauss of Krupp mentions the high resistance to rust and acids of the highly alloyed chromium-nickel steels in a lecture on microscopic examinations of steel at the main convention of German chemists in Bonn.

1914 **Firth's Aeroplane Steel.** During World War I, from 1914 to 1918, Firth's entire production of chromium stainless steel is requisitioned for use in aeroplane engine exhaust valves. The steel for this purpose is marketed under the name Firth's Aeroplane Steel, or FAS.

1915 **Brearley moves to Brown Bayley's.** On July 15, 1915, Harry Brearley, having resigned from Brown Firth Research Laboratories on December 27, 1914, becomes Works Manager at Brown Bayley's Steelworks, and the plant is soon making stainless steel.

1915 **Blades made by Mosely.** R.F. Mosely, Ltd. in Sheffield, England, becomes the second manufacturer of knives using Firth's iron-chromium alloy. The blades are stamped "Firth's Stainless Steel."

1915 **A nonrusting steel.** "Sheffield Invention Especially Good for Table Cutlery.

According to consul John M. Savage, who is stationed at Sheffield, England, a firm in that city has introduced a stainless steel, which is claimed to be non-rusting, non-stainable and non-tarnishable. This steel is especially adaptable to table cutlery, as the original polish is maintained after use, even when brought into contact with the most acid foods." *New York Times*, January 1915

1915 **Haynes files U.S. patent for chromium steel.** On March 12, 1915, Elwood Haynes of Kokomo, Ind., files a patent entitled "Wrought Metal Article." He says, "My invention relates to wrought metal articles of the nature of cutlery and edged tools that are incorrodible and lustrous, being composed of an iron-chromium alloy low in carbon and other metals, for example, an iron-chromium alloy containing no less than 8% chromium, and advantageously, not less than 10 or more than 50 to 60%, the best proportion of chromium being between 15 and 25%, and containing not to exceed 1% carbon (the amount of carbon being advantageously between 0.1 and 0.5% carbon) with the rest of the alloy consisting mainly of iron and chromium, there being no substantial amount (say not over 4 to 5%) of metals other than iron and chromium."

(The patent was at first denied, presumably because steels with chromium had already been patented. When Brearley received a patent for virtually the same material, it was contested by Haynes, who eventually was granted U.S. Patent 1,299,404 on April 1, 1919.)

1915 **Brearley Canadian patent.** On April 21, 1915, Harry Brearley of Sheffield, England, applied for and obtained Canadian Patent 164,122. Brearley claimed, "My invention relates to the production of steel or steel alloys and has for its object to produce a malleable steel which shall be practically untarnishable and can be forged, rolled, hardened and tempered under ordinary commercial conditions." Brearley claimed to have discovered "that the addition to iron of an amount of chromium between 9 and 16%, and also an amount of carbon not greater than 0.7%, will produce such a product." The patent was granted on August 31, 1915.

1915 **Advertisement.** An early advertisement for stainless steel in England reads as follows:

"Firth's 'STAINLESS' STEEL for CUTLERY, etc.
NEITHER RUSTS, STAINS NOR TARNISHES.
Cutlery made from this Steel, being totally unaffected by FOOD ACIDS, VINEGAR &c will be found a boon in EVERY HOUSEHOLD and may be had at all the LEADING MANUFACTURERS. The daily toil at the knife board or the cleaning machine is now quite unnecessary.
SEE THAT YOUR KNIVES OF THIS STEEL BEAR THE MARK **FIRTH/STAINLESS**
ORIGINAL AND SOLE MAKERS—
Thos. Firth & Sons, Ltd., Sheffield"

1915 **Firth-Sterling makes first stainless.** On March 3, 1915, the Firth-Sterling Steel Company of McKeesport, Pennsylvania, a subsidiary of Thomas Firth & Sons, Ltd., Sheffield, England, is the first American producer of stainless steel, a chromium martensitic alloy destined for an American knife manufacturer. The alloy was similar to type 420 stainless steel and was also known as cutlery steel.

1916 **Brearley patent.** On March 29, 1915, Harry Brearley of Sheffield, England, applies a continuation for his application for a

U.S. patent. He states, "My invention relates to new and useful improvements in cutlery or other hardened and polished objects of manufacture where non-staining properties are desired, and has for its object to produce a tempered steel cutlery or other hardened article having a polished surface and composition of an alloy which is practically untarnishable when hardened and tempered. The invention results from the discovery that the addition of certain percentages of chromium and carbon to iron will produce the characteristics above referred to." The patent covers alloys containing from 9 to 16% chromium and no more than 0.7% carbon and lists a typical alloy as having 13% chromium and 0.30% carbon, which is comparable to the modern type 420 alloy. Patent 1,197,236 was granted on September 5, 1916.

1917 **Stainless iron.** Stainless iron is developed at Firth-Sterling Steel Company, McKeesport, Pennsylvania, with a composition of approximately 13% chromium and 0.15% carbon. The alloy, which is ferritic, is not hardenable by heat treatment.

1917 **DIN.** Deutsches Institut fur Normung e.V. (DIN), the German standards organization, is established in Berlin.

1917 **Carpenter's first heats.** The Carpenter Steel Company, Reading, Pennsylvania, melts its first two heats of martensitic chromium stainless steel. The original uses are said to be in the Liberty airplane engine and for cutlery.

1917 **Firth-Brearley Syndicate.** The Firth-Brearley Syndicate is formed at Sheffield, England, to foster the worldwide use of stainless cutlery steel. The logo, "Firth-Brearley Stainless," was to be stamped on all knife blades made under the aegis of the syndicate. Firth's took an ingenious precaution to prevent companies from using the logo on other than Firth's stainless steel knives. When a melt of Firth's steel was ready for casting, a person from Firth's laboratory appeared with a small brown envelope, which was emptied into the molten metal. The envelope contained a small amount of element "X," the identity of which was known only to two persons in the company. Only a knife made of Firth's steel, when tested with a drop of a certain solution, would produce a pink spot. After a suitable number of years, the company revealed that element "X" was actually a small amount of cobalt.

1917 **Stainless steel.** According to the *Oxford English Dictionary*, the first published use of the term *stainless steel* was in the March 31, 1917, issue of *Scientific American Magazine*:

"The New Stainless Steel—A steel that does not stain nor tarnish is one of the latest new materials and will be welcomed by the housewife as a real boon. It is called 'stainless steel' and from it table cutlery is being made which not only takes a beautiful polish, but which preserves this appearance under all circumstances. It also keeps a sharp edge for a long time. Foods or ordinary acids do not tarnish it and it is fast becoming very popular. To be able to use ordinary knives or forks and by simply washing them, even without drying them, to have them maintain their original brightness is a much desired achievement and one eagerly welcomed by the diligent and particular housewife.

"The new steel was discovered in England but is now being made in the United States and sold as table cutlery. It is what is commonly called an alloy steel, that is, it differs from ordinary steel in that it contains a special element or metal. In this case it is chromium, which is mainly responsible for bestowing the stainless and rustless properties. By incorporating from twelve to thirteen per cent of this metal in mild carbon steel the new properties are obtained. An English metallurgist, in gathering some rods of steel which had lain a long time, noticed that while most of them were badly rusted or stained, a few were as bright as when originally made. This led to the present discovery and manufacture of the steel.

"The new steel is more expensive than the old which usually appears in the every-day table knife. Its advantages and lasting properties as well as its appearance and the convenience arising from its use more than offset the cost. It is supplanting not only the old steel cutlery but also the plated tableware. Its possibilities, however, are by no means limited to cutlery. One can readily imagine to what uses a stainless or rustless metal can be put."

(However, an earlier usage of *stainless steel* appeared in the *New York Times* in 1915.)

1917 **American Stainless Steel Co.** The American Stainless Steel Company, a patent-holding company, is organized in Pittsburgh, Pennsylvania. It is jointly held between the Firth-Brearley Syndicate (40%) and Elwood Haynes (30%), with the balance being shared between Firth-Sterling Steel Company, Bethlehem Steel Company, Carpenter Steel Company, Crucible Steel Company, and subsequently the Midvale Company.

According to K.C. Barraclough of the Firth Brown Research Laboratories:

"The formation of the American Stainless Steel Company proved to be a very satisfactory arrangement. Dividends of 25 percent were consistently paid during the late 1920s and early 1930s. It also ensured that the quality of stainless steel was kept under surveillance, with a flow of technical advice from Sheffield, backed up by occasional visits by Harry Brearley or by the directors of Firths and, eventually, the appointment of a technical manager who was trained in Sheffield and had access to such information as was required in America."

The American Stainless Steel Company was dissolved in the mid-1930s when the patent limits of Brearley, Haynes, and others had expired.

1919 **Haynes' discoveries.** Elwood Haynes, founder of the Haynes Stellite Company (succeeded by Cabot Corp. and Haynes International), discovered in 1911 that stainlessness could be imparted to steel by the addition of sufficient quantities of chromium. He immediately applied for a U.S. patent but was denied, apparently because chromium steels had previously been patented, despite the fact that the Haynes patent included the statement ". . . such surfaces are incorrodible." Harry Brearley had also applied for a U.S. patent having similar claims, and this was a further obstacle. Haynes protested and finally received U.S. Patent 1,299,404 on April 1, 1919.

Haynes and Brearley, in the author's opinion, are equal discoverers of the martensitic high-chromium stainless steels.

1919 **Ludlum Steel patent.** P.A.E. Armstrong of the Ludlum Steel Company, Watervliet, New York, receives a patent for an 8% chromium and 3% silicon steel. The alloy was widely used as a gasoline engine exhaust valve and sold under the name of Silchrome. The U.S. Patent number was 1,322,517.

The American Stainless Steel Company of Pittsburgh sued the Ludlum Steel Company for infringement of the Brearley and Haynes patents held by them, because these patents included a range of 8 to 60% chromium. After a long period of litigation, and a questionable decision by the court, the American Stainless Steel Company won the case.

1919 **Cutlery steel produced.** Following the end of World War I, during which all chromium stainless steel (Firth's Aeroplane Steel) in England was used for aeroplane valves, Sheffield cutlers start regular production of stainless steel cutlery, surgical scalpels, and tools. Early stainless steel tableware and bowls start to appear in hotels and restaurants in England.

1920 **Stainless iron.** The world's first commercial heat of stainless iron is cast at Brown Bayley's Steelworks, Ltd., Sheffield. In J.H.G. Monypenny's book, published in 1926, he states, "As far as the author is aware, such low carbon stainless steel (i.e., 0.1% carbon, or less), or stainless iron as it is quite generally called, was first produced on a commercial scale in June, 1920, when the firm with which he is associated made a five or six ton cast of material containing 0.07% carbon and 11.7% chromium and cast into twelve inch square ingots."

1920 **Latrobe produces stainless.** The Latrobe Steel Company begins producing stainless steel at Latrobe, Pa. They develop the first mirror-finish cutlery in the United States.

1920 **Carpenter No. 3.** The Carpenter Steel Company, Reading, Pennsylvania, introduces rustless steel, a ferritic chromium-copper alloy with a composition of 20% chromium, 1% copper, and 0.30% carbon. The alloy, known as Carpenter No. 3, was similar to the modern type 422. It was Carpenter's first stainless steel to be patented.

1920 **Sandvik produces stainless.** Sandvik Steel Company at Sandviken, Sweden, starts making stainless steel.

1920 **Haynes paper.** A paper, "Stellite and Stainless Steel," is presented by Elwood Haynes, president of the Haynes Stellite Company, at a meeting of the Engineers' Society of Western Pennsylvania. The paper describes Haynes' 1911 discovery of iron-chromium alloys that were found to be resistant to corrosion. Haynes applied for a U.S. patent in 1911, but it was at first denied because "patents for chromium steels had already been granted." Haynes was also contesting a patent application filed by Harry Brearley. Haynes eventually received his patent in 1920.

1920 **Marble paper on stainless.** The paper "Stainless Steel, Its Treatment, Properties and Applications" by W.H. Marble, manager at American Stainless Steel Company, Pittsburgh, along with a paper by Elwood Haynes, are among the earliest on stainless steel known to be published in America. Like all of the earliest American papers, it covers only the hardenable chromium stainless steels.

1920 **Maurer and Strauss paper.** The work on stainless steel by Eduard Maurer and Dr. Benno Strauss from 1909 to 1912 is published in *Krupp's Monthly* magazine in August 1920. It is of particular interest to note that the paper reveals, for the first time, the "Strauss-Maurer Chromium-Nickel Phase Diagram," which shows four crystalline phases.

Although patents had been applied for, this is the first time the details of the work are published. Dr. Strauss presented a paper on the steel at a symposium of the American Society for Testing Materials in 1924.

(This delay in distributing technical information on the austenitic chromium-nickel stainless steel may well be the reason that this steel was not used or made in North America until 1927.)

1920 **Brearley's Bessemer Medal.** Harry Brearley receives one of the highest awards for metallurgical achievement. The Bessemer Gold Medal is bestowed upon him by the British Iron and Steel Institute for his work on the discovery and commercialization of chromium stainless steels. The award was established and endowed by Sir Henry Bessemer of Sheffield. Harry Brearley received the fourth of the nine medals awarded between 1874 and 2008.

1921 **Victorinox knives.** A cutlery company was founded by Karl Elsener in the town of Ibach, in the Canton of Schwyz, Switzerland, in 1884. The company began delivering knives to the Swiss Army in 1891. In 1909, upon the death of his mother, Victoria, Elsener changed the name of his company to Victoria. In 1921, when Elsener started to make knife blades of stainless steel, he changed the company name to Victorinox, a combination of his mother's name and "inox" for *inoxidable*, the French word for stainless. Since 2005, Victorinox has been the sole producer of Swiss Army knives.

1922 **Low-carbon stainless.** The General Electric Company at Schenectady, New York, makes its first heat of low-carbon, 12% chromium ferritic stainless steel.

1922 **Nirosta.** Nirosta is a trade mark or brand name registered by Krupp. The name applies to a variety of stainless steels produced by Krupp. It is an acronym for *nicht-rostender-Stahl*, or nonrusting steel.

1922 **Boiler tubes.** Chromium stainless steel with 0.30% carbon and

12% chromium is fabricated into boiler tubes by Babcock and Wilcox Tube Company, Beaver Falls, Pennsylvania.

1923 **First stainless book.** The world's first book in the English language on stainless steel is published in 1923 by the Firth-Sterling Steel Company at McKeesport, Pennsylvania. They were the first producers of martensitic chromium stainless steels in America, in 1915, and the first producers of ferritic chromium steels in America, in approximately 1920. Until about 1927, these alloys were the only stainless steels used in America.

Firth-Sterling 'S-Less' Stainless Steel is a 70-page book that contains a brief history of stainless steel and information on five types of chromium stainless steels, including physical and mechanical properties, heat treating and working, and the corrosion resistance properties with regard to 68 substances.

The book preceded Monypenny's book, printed in London, by three years. (See also 1926, Monypenny book.)

1923 **Super stainless steels.** In approximately 1923, when the austenitic chromium-nickel stainless steels are introduced into the United Kingdom from Germany, they are called super stainless steels in England to distinguish them from the plain chromium stainless steels.

1923 **Budd introduces an all-steel car body.** The Edward G. Budd Manufacturing Company of Philadelphia introduces an all-enclosed steel car body that has a rounded, more attractive shape than the standard boxlike appearance of current models.

1923 **Firth-Brearley Syndicate and Krupp exchange licenses.** The Firth-Brearley Syndicate comes to an agreement with Krupp whereby licenses for 13% chromium steel will be exchanged for permission to make Krupp's austenitic steels in England.

This was the catalyst for an enormous effort directed toward the investigation of austenitic corrosion-resistant materials at the Brown Firth Research Laboratories and elsewhere in Sheffield.

1924 **Hatfield invents 18-8.** Dr. William A. Hatfield, who was Harry Brearley's successor at the Brown Firth Research Laboratories in Sheffield, is credited with the invention of 18-8 stainless steel (18% chromium and 8% nickel). This is a modification of the iron-chromium nickel alloy discovered by Strauss and Maurer, which was a 20% chromium and 7% nickel alloy. Hatfield's alloy became the preferred alloy and even the most popular stainless alloy to this day.

Dr. Hatfield also invented 18-8 stainless steel with the addition of titanium to prevent chromium carbide precipitation, an alloy now known as type 321. The titanium combines with carbon to form chromium carbide, leaving the protective chromium in solution.

Hatfield is said to have remarked that over 15,000 tests had been made in his laboratories to determine the resistance of stainless steels to all strengths of acids and reactants likely to be met in industrial processes, as well as to some domestic items, such as tomato catsup, curry sauce, and even whiskey.

1924 **Parmiter paper on cutlery steel.** Owen Parmiter of Firth-Sterling Steel Company, McKeesport, Pennsylvania, presents what is probably the first detailed American paper on Brearley's cutlery stainless steel at a meeting of the American Society for Steel Treating in September 1924. It is also the earliest paper known to include micrographs of stainless steel and photographs of stainless steel tableware. The full paper was reprinted in Thum's book on stainless steels, published in 1933.

1924 **Avesta produces chromium steel.** Avesta Jernverk produces chromium stainless steel in Sweden.

1924 **Allegheny Steel's first patent.** Allegheny Steel and Iron obtains its first stainless steel patent.

1924 **Sandvik's chromium stainless tubes.** The first seamless chromium alloy stainless steel tubes are produced by the Sandvik Steel Company, Sandviken, Sweden.

1924 **ASTM Symposium.** The first large symposium concerning stainless steel is held by the American Society for Testing Materials at Atlantic City, New Jersey. The "Heat- and Corrosion-Resisting Alloys and Electrical-Resistance Alloys" symposium included seven papers on stainless steels, one of which consisted of three large charts that showed data on stainless steels produced by ten companies, including chemical compositions, physical and mechanical properties, and corrosion data.

1924 **Strauss presents paper.** Dr. Benno Strauss of the Krupp Works presents a paper, "Non-Rusting Iron-Chromium-Nickel Alloys," at the ASTM Symposium in Atlantic City. Although the paper describes the work done by him and Eduard Maurer in the period from 1909 to1912, this is apparently the first time the details of the work are presented to the public.

Dr. Strauss discussed the chemical compositions and the properties of the V1M and V2A alloys, for which Krupp attained German patents in 1912. The compositions of these alloys were as follows:

	V1M (martensitic), %	V2A (austenitic), %
Carbon	0.22	0.25
Silicon	0.07	0.08
Manganese	0.12	0.11
Nickel	1.75	7.00
Chromium	10.05	20.10

(Note: The austenitic alloy had not yet been used or produced in the United States, but Strauss' paper must have been of immense interest to the American producers present as well as to men working at firms needing corrosion-resistant metals.)

1924 **Armstrong paper on history.** P.A.E. Armstrong, vice president of Ludlum Steel, Watervliet, New York, presents the third paper at the ASTM Symposium in Atlantic City, New Jersey, on "Corrosion- and Heat-Resisting Alloys." It is an excellent 13-page history of stainless steels and other metals and includes references to and discussions of a great many stainless steel patents.

1924 **Stainless tanks.** The first application of stainless steel plate for tanks takes place when a large chemical plant in the United Kingdom installs stainless steel tanks for the storage of nitric acid.

1924 **Struthers Wells fabricates 18-8.** P.F. McEvoy at the Struthers Wells Company in Warren, Pennsylvania, states that 18-8 stainless steel is being fabricated in commercial unit size. (This is the earliest known date for the fabrication of 18-8 stainless steel in the United States. The source of the steel was not given.)

1925 **Ferrochromium specification.** ASTM specification A 101, "Ferrochromium," is published.

1925 **Avesta produces austenitic stainless.** Avesta Jernverk produces the first austenitic stainless steel in Sweden.

1926 **Monypenny book.** The world's second book on stainless steel is published in London. John Henry Gill Monypenny (1885–1949), Chief of the Research Laboratory at Brown Bayley's Steel Works, Ltd., Sheffield, is the author of *Stainless Iron and Steel*. The well-illustrated 300-page volume includes a history of stainless steel and provides a thorough discussion of the characteristics, properties, and applications of stainless iron (ferritic iron chromium), stainless steel (martensitic iron-chromium steel), and austenitic stainless steel (iron-chromium-nickel alloys). Micrographs, graphs, and tables are included.

(See also 1923, first stainless book for information on the world's first book on stainless steel, published by the Firth-Sterling Steel Company of McKeesport, Pa., in 1923).

"In the application of stainless steel the microstructure is of especial value as the observations indicate not only the cause of varying mechanical properties but also reasons why the materials offer varying resistance to corrosion."

J.H.G. Monypenny, 1933

1926 **Surgical implants.** A stainless steel having 18% chromium and 8% nickel is introduced into surgical implant applications. The material is noted to be much more resistant to bodily fluids and stronger than the vanadium steel introduced by Sherman for his fracture-fixation plates.

1926 **Allegheny Steel metallurgists visit Hatfield at Sheffield.** Allegheny Steel Company sends a delegation to visit William H. Hatfield, head of the Brown Firth Laboratories at Sheffield, to learn about the 18-8 austenitic stainless steel that Hatfield had recently invented. It is almost identical to the modern type 304 and is a modification of the Krupp alloy.

1926 **Rustless Iron Company opens in Baltimore.** The Rustless Iron Company (not to be confused with the Rustless Iron & Steel Co. established in 1930) is established in Baltimore, Maryland, by Alwyn and Ronald Wild from Sheffield, England. The Wilds have a new process for making stainless iron that uses chromite ore instead of ferrochromium for making stainless steel. The Wilds were unable to produce rustless iron with a low enough iron content, and the company went bankrupt in 1930.

1926 **Rustless process invented by A.L. Feild.** Working along lines similar to that of the Wilds, Feild developed a process for making low-carbon stainless steel that was much less expensive than when using low-carbon ferrochromium.

1926 **Budd Wheel Company produces dual wheels**. The Budd Wheel Company in Detroit, a subsidiary of the E.G. Budd Manufacturing Company, introduces dual wheels for trucks and buses. In a joint venture with Michelin Tire of France, Budd produces steel dual wheels with pneumatic tires furnished by Michelin, so there are four tires on each axle, for greater load-carrying ability.

The arrangement with Michelin will prove to be very fortunate for the Budd Company.

1927 **First austenitic steel article.** The first known paper on austenitic stainless steel, entitled "An Introduction to Iron-Nickel-Chromium Alloys," is published in America by E.C. Bain and W.E. Griffiths of the U.S. Steel Company Research Laboratory at Kearny, New Jersey. It appeared in *Trans. AIME*, Vol 117, 1927, p 383.

1927 **Carpenter Steel forms tube division.** Carpenter Steel Company forms the Welded Alloy Tube Division at Kenilworth, New Jersey, to make tube produced on their new strip mill at Reading, Pennsylvania.

1927 **Hatfield introduces DDQ alloy.** Dr. William H. Hatfield of the Brown Firth Research Laboratories develops a deep-drawing-quality (DDQ) alloy with 12% chromium and 12% nickel that has improved ductility over regular 18-8 and is reliably nonmagnetic.

1927 **Krupp patent on 25-20.** Krupp obtains a German patent for an austenitic 25% chromium and 20% nickel steel, which subsequently becomes known as 25-20 and type 310 stainless steel.

1927 **Stainless cookware.** The first stainless steel cookware is produced by the Polar Ware Company, Sheboygan, Wisconsin, primarily for commercial use.

1927 **Leipzig Fair.** Visitors to the Leipzig Spring Fair see distilling apparatuses, acid pumps, turbine blades, beer barrels, tableware, and kitchenware made of Krupp Nirosta (*nicht-rostender Stahl*, or nonrusting steel). This was an austenitic stainless steel of the 18-8 type.

1927 **Heil Truck produces first welded milk truck.** Heil Truck of Milwaukee, Wis., produces a welded stainless steel tank of chromium stainless steel. Heil became one of the largest producers of trucks for refuse collection and has 200 patents.

1927 **4 to 6% chromium tubes.** In November 1927, the first 4 to 6% chromium steel heater tubes were installed in an American oil refinery for handling hydrogen sulfides. This steel, although corrosion resistant, was not stainless steel, but, as an expedient, AISI later assigned the 500 series of stainless steel numbers to the 4 to 6% chromium steels.

1928 **Hatfield visits Allegheny Steel.** William H. Hatfield, head of the Brown Firth Laboratories, Sheffield, visits the Allegheny Steel Company, Watervliet, New York, and "found a considerable production of 18-8 stainless steel."

1928 **Carpenter No. 5.** Carpenter Steel Company, Reading, Pennsylvania, introduces Carpenter No. 5, an antifriction stainless steel known today as type 416. It is a straight chromium grade with the addition of at least 0.15% sulfur to make it easier to machine. It was the world's first free-machining stainless steel.

1929 **Precipitation hardening discovered.** William J. Kroll (1889–1973) of Luxembourg is the first to discover precipitation-hardening stainless steel. He used titanium. Kroll developed the Kroll process for refining titanium and zirconium.

1929 **Hotel Savoy.** The first recorded use of stainless steel in architecture is a large stainless steel sign and sidewalk canopy erected at the entry to the Hotel Savoy in London.

1929 **Ford cars with stainless trim.** The Ford Motor Company starts using Allegheny Metal stainless steel for the bright trim of the Model A Ford car.

1929 **Pierce Arrow has 24 pounds of stainless.** Pierce Arrow uses 24 pounds of Carpenter stainless steel strip as trim on each car.

1929 **Milk tanker.** A 3000 gallon milk tanker is the first stainless steel rolling stock.

1929 **Stainless steel golf clubs are manufactured.** Stainless steel golf clubs are manufactured for the first time.

1929 **ASTM Committee A-10 on Stainless Steel.** The American Society for Testing Materials (ASTM) at Philadelphia establishes Committee A-10 on Corrosion- and Heat-Resisting Stainless Steels. Jerome Strauss, Chief Research Engineer at the Vanadium Corporation, is the organizer and becomes the first secretary of the committee.

1929 **Feild produces rustless iron.** A.L. Feild produces rustless iron commercially in a 6 ton electric furnace at Lockport, N.Y.

1929 **U.S. stainless steel production.** T.J. Lipper reports in the January 31, 1931, edition of *Iron Age* that the total stainless steel production in 1929 was 53,293 tons.

1930 **Oxide films.** U.R. Evans (1889–1980), Cambridge, England, provides direct evidence of the invisible oxide films on stainless steel by a simple electrochemical technique that makes it possible to peel the film from a passive surface and view it under a microscope. It is the chromium oxide film that is responsible for the corrosion resistance of stainless steels.

1930 **Chrysler Building.** The 77-story, 1046 foot high Chrysler Building, the tallest structure in the world, is erected in New York City. Walter Chrysler, president of the Chrysler Motor Company, wanted the building for his company headquarters to be not only the tallest skyscraper but also the finest and most ornate.

He selected Nirosta 20-7 stainless steel to adorn the 100 foot spire of the building, as well as for the enormous decorative pineapples, gargoyles, and giant Chrysler radiator caps, each of which had a wing span of 15 feet, as well as for all roofing, sidings, flashing, and coping. Altogether, there was approximately 48 tons of the German-patented Nirosta metal, making the Chrysler Building the world's first major use of stainless steel in architecture.

No one could predict how long this metal would last in New York's polluted atmosphere near the sea, but after almost 80 years, the spire is still silvery bright.

The building has become the epitome of stainlessness. Photographs of the Chrysler stainless steel spire have appeared prominently in recent years in brochures published by A-K Steel, Allegheny Ludlum, and KruppThyssen Nirosta.

1930 **Rustless Iron and Steel Company founded in Baltimore.** A.L. Feild starts the Rustless Iron and Steel Company in Baltimore, Maryland, in the plant vacated by the Wild brothers. The new company will make low-carbon rustless iron.

1930 **The Huey corrosion test.** In approximately 1930, William R. Huey, a corrosion consultant in Swarthmore, Pennsylvania, develops a corrosion test for chromium stainless steel that was used by the E.I. DuPont de Nemours Company. In the test, thin metal specimens approximately 1 inch in diameter are suspended from glass hooks in a closed flask containing approximately 750 cubic centimeters of 65% boiling nitric acid solution. The specimen is examined and weighed at five 48 hour intervals, each period starting with fresh acid. The weight of the metal lost is converted into inches per year.

This is ASTM Method A 262, Practice C.

1930 **Clad plate introduced by Lukens Steel.** A process for roll bonding metals such as stainless steel and nickel to a backing plate of carbon steel is developed by the Lukens Steel Company in Coatesville, Pennsylvania, working with the International Nickel Company of New York City. The process was developed to reduce the cost of heavy plates that needed to be corrosion resistant on just one side. Typically, the cladding material is 1/10 the thickness of the backing material. A metallurgical bond is formed during the hot rolling process.

1930 **U.S. stainless steel production.** T.J. Lipper reports in the January 31, 1931, edition of *Iron Age* that the total stainless steel production in 1930 was 26,618 tons.

1931 **Molybdenum in stainless alloy.** The first mention of molybdenum-bearing 18-8 stainless steel in the literature appears to be in a paper by W.H. Hatfield and H.T. Shirley in the first report of the ISI Committee in 1931.

1931 **Empire State Building is trimmed with stainless.** Following the precedent set by the Chrysler Building a year earlier, the Empire State Building has stainless steel window trim and pilasters. Krupp Nirosta 18-8 chromium-nickel steel strip and sheet was supplied by the Republic Steel Company and the Allegheny Steel Company. The 102-story building was the tallest in the world for 40 years.

1931 **Low-carbon austenitic stainless.** The world's first very low-carbon (0.02%) austenitic stainless steel is produced by Aciéries d'Unieux (later integrated into the Compagnie des Ateliers et Forges de la Loire) in France.

1931 **First stainless aircraft.** The world's first stainless steel aircraft, the Pioneer, is built by the Edward G. Budd Manufacturing Company in Philadelphia. Plans for the Italian Savoia-Marchetti were obtained in order to construct the plane using 0.006 inch thick stainless steel sheet purchased from the Allegheny Steel Company at Brackenridge, Pennsylvania. The plane, at 1750 pounds, was a little lighter than the Italian wooden planes.

It was test flown in 1932. After demonstration flights in America, the Pioneer I was shipped to Europe, where demonstration flights were made in England, France, and Italy. After logging 1000 hours of flight, the plane was brought back to America and donated to the Franklin Institute in Philadelphia, where it is on display outside the museum.

The plane, a four-passenger, two-wing amphibian, was built using very thin 18-8 austenitic chromium-nickel stainless steel sheet for its hull and frame, cold rolled to a strength of 150,000 pounds per square inch. Building an airplane was a bit of showmanship on the part of Mr. Budd, who simply wanted to show the world what the Budd Company could do with this new material that was only recently introduced to America. (In an early advertisement, Budd had shown, by use of trick photography, an elephant standing on the roof of a Budd-built, welded steel automobile body.)

Budd demonstrated that he could fabricate the metal into an airplane that was light enough to fly. Engineers at Budd had also learned how to form and weld the new material. The stunt paid off. The experience gained by working with stainless steel led Budd to manufacture hundreds of stainless steel passenger trains and subway cars over the next 55 years.

1931 **Shot weld patent.** Col. Earl J.W. Ragsdale, research engineer at the Edward G. Budd Manufacturing Company, Philadelphia, applies for a U.S. patent relating to the electric resistance spot welding of cold-rolled 18-8 stainless steel. The application explains how cold-rolled stainless steels having tensile strengths of up to 250,000 pounds per square inch can be spot welded without reducing the strength of the metal in the area surrounding the weld and without permitting the formation of carbides, or annealing, in the area around the weld. This is accomplished by using very high currents and very short welding times, on the order of one-hundredth of a second.

This process permitted, for the first time, the welding of high-strength stainless steel structures without degrading the strength or reducing the corrosion resistance at the welds. A patent was granted to Ragsdale on January 16, 1934. The process was used extensively by the Budd Company in the building of stainless steel trains, cargo planes, and other structures. The Budd patent prevented other companies from entering this field of business for many years.

1931 **Second edition of Monypenny's book.** The world's third book and America's second book on stainless steel is published by John Wiley & Sons, New York. It is the second edition of *Stainless Iron and Steel* by J.H.G. Monypenny. The 575-page volume is almost twice the size of the first edition published in London in 1926. Monypenny went on to publish six editions of his book.

1931 **Plummer system for grouping stainless steels.** A system listing stainless steel alloys in groups according to their alloy content is developed by Clayton E. Plummer of the Robert W. Hunt Company, Chicago. It is published in the July 1931 edition of *Factory and Industrial Management.* Plummer listed 919 alloys made by 98 producers in 29 groups in accordance with percentage ranges for carbon, chromium, and nickel. This was approximately the same time that the AISI numbering system was set up for stainless steels.

1932 **Budd stainless trailer.** The Edward G. Budd Company of Philadelphia inaugurates lightweight stainless steel trailer body production using high-strength, cold-rolled type 301 stainless steel sheet attached to a stainless steel frame by the patented Budd shot weld process.

1932 **Budd-Michelin rubber-tired train.** The Edward G. Budd Company of Philadelphia designs and builds a self-propelled, rubber-tired, stainless steel passenger car. At a meeting with Mr. Budd in Paris in September 1931, M. Hauvette of the Michelin tire company disclosed a wheel with a flange that would permit use on a railroad track. He claimed that this invention would revolutionize railcars, giving passengers a smoother and quieter ride. Hauvette offered to exchange rights to the patented wheel for rights to use the Budd Company's shot weld patent. Budd accepted the offer and proceeded to design and build a railcar that would be light enough to use pneumatic tires.

The roof and siding of the car are made of thin, very high-strength (150,000 pounds per square inch) 18-8 stainless steel sheet that is attached to the stainless steel frame by a patented electric resistance welding process called the Budd shot weld process. The shot weld process was the only welding process that could produce spot welds in cold-rolled stainless steel sheets without causing a loss of corrosion resistance and strength in the area around the weld.

The rubber-tired car, which was affectionately called the "Green Goose," weighed only approximately half that of a car made of carbon steel. Three other Budd-Michelin rubber-tired trains were built and sold to the Pennsylvania Railroad, the Reading Company, and the Texas & Pacific Company. The Reading train, Car No. 65, served as a commuter train on the ten-mile run between the towns of New Hope and Hatboro, Pennsylvania, for 14 years.

The rubber-tired trains were a failure, because the tires did not hold up well and slipped on wet rails. However, the trains were the forerunners of fleets of streamlined stainless steel trains with steel wheels, which Budd would sell to 80 railroads over the next 55 years. (See also 1934, Burlington Zephyr.)

1932 **Rolls-Royce engine with stainless parts.** The Supermarine SB6 gives the Royal Air Force its third consecutive win in the Schneider Trophy race for seaplanes. Its Rolls-Royce aero-engine had stainless steel shafts, rods, valves, and spindles.

1932 **Patent for stainless railcar body.** Earl J.W. Ragsdale and Albert G. Dean of the Edward G. Budd Manufacturing Company, Philadelphia, file a patent application on April 22, 1932, for a stainless steel railcar body. U.S. Patent 2,129,235 was granted on September 6, 1938.

1933 **Thum book on stainless steels.** The second book on stainless steel written in the United States, *The Book of Stainless Steels*, is published by the American Society for Metals, Cleveland, Ohio. The 700-page book is edited by Ernest Thum, Editor of *Metal Progress* magazine, who called on 75 experts to produce the book's 75 chapters. The book covered every possible phase of manufacturing and fabrication and addressed dozens of applications. It included a list of the recently established AISI numbers of stainless steel and their chemical compositions, but these numbers were not used by the various authors. A second edition was published in 1935.

1933 **Alloy naming**. "Probably most attentive readers of these first few pages have sensed the difficulties which arise in the classification and nomenclature of the high-chromium alloys. Especially is the latter badly cluttered up with trade names and loose terminology." Ernest Thum, Editor, *The Book of Stainless Steels*, 1933

1933 **Intergranular corrosion.** E. Houdremont (Fig. 15) and P. Schafmeister (Fig. 15) publish an early paper, "Prevention of Intergranular Corrosion in Steels with 18% Chromium and 8% Nickel with the Addition of Carbon-Forming Metals," in *Archiv für das Eisenhüttenwesen*, Vol 7, 1933–1934, p 187.

1933 **Trade names.** By 1933, there were 90 North American companies producing or fabricating stainless steel and marketing the products by their trade names.

1933 **U.S. Steel Corp. exhibits kitchen ware at Chicago Exposition of Progress.** A large showcase at the Chicago Centennial Exposition of Progress displays various items made of the 18-8 stainless alloy, including a kitchen sink, countertop, pots, and pans.

1933 **Duplex stainless steel discovered.** Avesta Ironworks in Sweden develops the first stainless alloys that have microstructures consisting of ferrite and austenite. Known as duplex alloys, they have considerably higher strength and better resistance to stress-corrosion cracking than either the ferritic or austenitic alloys. One early duplex alloy became AISI type 329. The alloys did not become popular until the advent of the argon-oxygen refining process, which was introduced in the 1970s.

1933 **AISI numbers for stainless steel.** In approximately 1933, the Stainless Steel Committee of AISI developed a numbering system for wrought stainless steels. Such a system was badly needed, because, by that date, there were approximately 90 companies in the United States and Canada that were marketing stainless steels and naming them with approximately 1000 different trade names, not to mention the trade names used in Europe.

A three-digit system was selected, because AISI had already established a four-digit numbering system for carbon and alloy steels. The system consisted of a 3*xx* series of numbers for the iron-chromium-nickel austenitic alloys, a 4*xx* series for both the ferritic and martensitic iron-chromium stainless steels, and a 5*xx* series for 4 to 6% chromium alloys. This last group was included only because these alloys had sufficient corrosion resistance to be useful in the petroleum refining industry, but they did not fit into the AISI four-digit numbering scheme for alloy steels.

Unfortunately, the acceptance of 4% chromium steels in the AISI numbering system caused the definition of stainless steels to appear in dictionaries for many years as "steels containing at least four percent chromium."

The AISI had certain rules for assigning numbers that included the production of at least 1000 tons of an alloy per year. The AISI discontinued the assignment of numbers in approximately 1960. Those stainless steels that had received AISI designations became known as the "standard" stainless steels. All other stainless steels in published lists were classified as "nonstandard" or "other" stainless steels. The AISI numbering system for stainless steels remains, along with the German DIN Werkstoff Numbers, as the world's most widely used designation system for carbon, alloy, and stainless steels.

1934 **Union Pacific M-10000 is dedicated.** In February, this Pullman-built, three-car, all-aluminum articulated train is the first streamliner in the United States. It is powered by a Winton V12 600-horsepower distillate engine and is capable of 110 miles per hour. It made a 12,625 mile coast-to-coast exhibition trip and was seen by almost 1.2 million people at various stops. The train went into service as the City of Salinas on January 31, 1935. In 1942, the train was scrapped because of the aluminum shortage during World War II.

1934 **Rustless stainless melting process.** The Rustless Iron and Steel Company, Baltimore, Maryland, develops the Rustless stainless steel melting process, which is the first use of stainless scrap and chromite ore, instead of ferrochrome, to make stainless steel at a reduced cost.

1934 **Burlington Zephyr's record run.** The three-car Burlington Zephyr, a streamlined, lightweight stainless steel train built by the Budd Manufacturing Company at Philadelphia, makes its first scheduled run on May 24 from Denver to Chicago. The train was purchased by the Chicago, Burlington, & Quincy Railroad in an effort to stimulate passenger travel, which had fallen greatly during the midst of the Great Depression. The body of the car is made of type 301 stainless steel, cold rolled to 150,000 pounds per square inch minimum tensile strength. The structure is assembled by the patented Budd shot weld method. The high-strength steel had never been used before the development of this welding method.

The three-car train, which weighs 104 tons, including 23 tons of stainless steel, is approximately the weight of a single Pullman coach. The train is sleek, shiny, and powered by one of the newly developed diesel-electric Winton engines. The interior features comfortable and stylish modern furnishings as well as air conditioning. The train, with 72 passengers and mail on board, makes the nonstop, dawn-to-dusk, 1014 mile trip in a little over 13 hours, cutting in half the time for the normal steam train run. Fuel consumption for the trip was 419 gallons and cost just $16.72, with the cost of diesel fuel at four cents per gallon. In other words, thanks to the streamlining, the light weight, and a remarkably efficient diesel engine, the 100 ton train got two miles per gallon and cost 1.6 cents per mile.

The Burlington Zephyr represented just the first of what would become hundreds of Budd-built stainless steel trains for 80 American and some foreign railroads in the coming years.

Following its record-breaking run to Chicago, the train was exhibited for six months.

1934 **NEW TRAIN SMASHES WORLD SPEED MARKS**

Burlington Zephyr Eclipses Royal Scot's
Record in Non-stop Run Denver to Chicago;
Averages 77.5 M.P.H; Top of 112.5

New York Times, May 26, 1934

1934 **Exhibit of Burlington Zephyr.** Following its record-breaking run to Chicago, the train was exhibited for six months at the Chicago's Century of Progress Fair that was then in progress. Next, the train was taken on a nationwide tour and displayed at 222 cites. The train was in regular service for 26 years, having covered 3.2 million miles. The train, which has been refurbished, is on permanent display at the Chicago Museum of Industry and Science.

1934 **Ragsdale receives patent for shot welding.** Earl J.W. Ragsdale, Chief Engineer of the E.G. Budd Manufacturing Company, Philadelphia, receives a patent for a method of spot welding cold-rolled austenitic 18-8 stainless steel sheets without reducing the strength or affecting the corrosion resistance of the base metal adjacent to the weld. It was an automatic electric resistance welding method called shot welding, wherein the welding time was only a fraction of a second—too short a time to adversely affect the properties of the cold-rolled steel.

The discovery of this welding process was a critical factor in the building of the Budd stainless steel trains, truck trailers, and cargo planes.

1934 **"The Silver Streak" film.** While at the Chicago Fair in 1934, the Burlington Zephyr, also known as Train No. 9900, was borrowed by Hollywood to become the star of an RKO movie called "The Silver Streak." In the movie, the President's son contracts polio and is in immediate need of an iron lung. An iron lung is loaded onto the Silver Streak, which speeds across the continent, delivering it in time to save the President's son.

1934 **Philadelphia skyscraper is adorned with stainless.** The third skyscraper to be built with considerable amounts of stainless steel for outside trim and interior ornamentation is the Philadelphia Savings Fund Society Building.

1934 **Carpenter patent for type 303.** Carpenter Steel Company is granted a patent for their No. 8 alloy, the world's first free-machining chromium-nickel stainless steel. It is known today as type 303. It is a type 302 alloy containing a minimum of 0.15% sulfur.

1934 **American production of stainless.** The total American production of corrosion- and heat-resisting alloys, including castings, for 1934 was as follows:

Type	Tons
18-8	25,733
12–14% Cr	8,822
16–18% Cr	6,328
All others	5,680
	46,563

T.J. Lipper, in the January 31, 1935, edition of *Iron Age*

1935 **SAE develops stainless numbering system.** In approximately 1935, the Society of Automotive Engineers (SAE), following the lead of AISI, develops its own numbering system for stainless steels. In general, the AISI system is adopted, with the exception that two digits precede the three-digit AISI numbers. The digits "30" precede the numbers for the 3*xx* austenitic stainless steel series (e.g., 30304); the digits "51" precede the numbers for the 4*xx* chromium stainless steel series (e.g., 51410) and the 5*xx* series for the 4 to 6% chromium alloys (e.g., 51501).

1935 **Stainless Fords built.** Officials at Allegheny Ludlum and the Ford Motor Company collaborate on an experiment that will become a legacy and a tribute to stainless steel. Six Ford Deluxe sedans are manufactured for display by Allegheny Ludlum to increase interest in the use of stainless steel.

1935 **Mark Twain Zephyr christened.** The third Budd-built stainless steel train is christened at Hannibal, Missouri. It is a four-car, self-propelled train and is the second of what was called "the Burlington's gleaming symphony of 18-8 stainless steel, the legendary Zephyrs."

1935 **Alloy Casting Industry classifies stainless alloys.** The Alloy Casting Industry Code Authority puts stainless casting alloys into three classifications: "C" for chromium alloys, "CN" for chromium-nickel alloys, and "NC" for nickel-chromium alloys. The letters are followed by one or two digits, such as C20,

CN35, and NC5. (This should not be confused with the Alloy Casting Institute. See 1940.)

1935 **Dupont alloy 20**. Dr. Mars A. Fontana, working at the Experimental Station of the E.I. DuPont de Nemours Company in Wilmington, Delaware, develops what becomes known as alloy 20, which is the first stainless steel suitable for handling sulfuric acid. It is an iron-base casting that contains 20% chromium, 20% nickel, 2.25% molybdenum, and 3.25% copper. Other designations for the cast alloy 20 are ACI CN-7M and UNS N08007.

1935 **First ASTM specifications for wrought stainless steels.** The first ASTM specifications for stainless steel mill products are published: ASTM A 167, "Stainless and Heat-Resisting Chromium-Nickel Plate, Sheet and Strip," and ASTM A 176, "Stainless and Heat-Resisting Chromium Plate, Sheet and Strip." These are the first specifications to be written by ASTM Committee A-10 on Stainless Steel, which was established in 1929.

The alloys in these specifications are designated according to the new AISI three-digit designations, such as 304, 310, 410, and 430.

1935 **First ASTM specifications for cast stainless steels.** In 1935 and 1936, stainless steel casting alloys start to appear in ASTM specifications. Eight specifications, from A 168 through A 175, describe the alloys, which are identified only by their chemical compositions. Alloy Casting Institute (ACI) designations would not be developed until 1941.

1935 **The Flying Yankee arrives at Boston.** "Self-propelled, stainless steel articulated unit will meet fast schedules on the Boston & Maine and Maine Central between Boston, Portland, Me., and Bangor."

On February 9, 1935, the train known as the Flying Yankee is delivered by the E.G. Budd Manufacturing Co., Philadelphia. It is the first of its kind to be delivered to an Eastern railroad. The three-car train is very similar to the Burlington Zephyr that was placed in service in 1934 by the Chicago, Burlington, & Quincy Railroad.

1935 **Stainless household sinks.** In approximately 1935, sinks made of 18-8 stainless steel begin to be installed in new homes, instead of the heavy porcelain-enameled cast iron sinks.

1936 **Italian firm licensed to manufacture Budd railcars.** The first stainless steel railcars to be built in Europe are produced at the

Piagio Works in Genoa, Italy, under license from the E.G. Budd Manufacturing Co., Philadelphia.

1936 **G.O. Carlson, Inc. to process plate.** Gunard O. Carlson opens a shop at Downingtown, Pennsylvania, to fill a need for stainless steel discs, rings, heads, tube sheets, and special-cut shapes. The stainless steel plate will generally be furnished by the nearby Lukens Steel Company at Coatesville, Pennsylvania. The business was started primarily to fill the need for stainless steel chemical equipment at the nearby E.I. duPont de Nemours Co. in Wilmington, Delaware.

1936 **Pullman Standard applies stainless sheathing on passenger cars.** Stainless steel is first applied by Pullman Standard as an exterior sheathing on passenger cars in an effort to compete with Budd trains.

1936 ***Queen Mary's* stainless applications.** On September 26, the RMS *Queen Mary* is launched at Southampton and will reach New York City in 4½ days. Stainless steel is widely used throughout the vessel in its kitchens, swimming pools, interior décor, and turbine engines.

1937 **Budd receives order for 104 cars.** The Budd company receives its first large order for stainless steel railcars, an order for 104 passenger cars for the Atchison, Topeka, & Santa Fe Railroad's new train, The Super Chief. The order is worth over $5 million and will require approximately 800 tons of type 301 stainless steel.

1937 **Crucible patent for stabilized stainless.** The Crucible Steel Company of America receives a patent for the stabilization of austenitic stainless steel to avoid carbide precipitation.

1937 **Alloy Casting Research Group organized.** A group of major producers form to conduct studies on the heat- and corrosion-resistant stainless steel casting alloys to determine baseline properties and the influence of composition variations. This group led to the organization of the Alloy Casting Institute. (See 1940).

1938 **Allegheny Steel and Ludlum Steel merge.** The Allegheny Ludlum Steel Corporation is created by a merger of Allegheny Steel Company, Pittsburgh, and Ludlum Steel Company, Watervliet, New York.

1938 **ASTM specification on castings is published.** ASTM A 219 on martensitic stainless steel castings is published.

1938 **Steel Founders' Society begins technical activities.** With the appointment of Charles W. Briggs, the Steel Founders' Society of America (SFSA) begins to address technical issues.

1938 **ASTM specification on boiler tubes is published.** ASTM A 213, "Seamless Ferritic and Austenitic Boiler Tubes," is published.

1938 **Nuts and bolts.** ASTM A 194 on nuts and A 193 on bolts are published. These are the first ASTM specifications for fasteners to cover stainless steels.

1938 **ASTM specification on sanitary tubing.** ASTM A 270 on sanitary austenitic stainless steel tubing is published.

1938 **Collection of Oscar Bach's stainless steel metalwork**. A door, with its trim and grillwork, is placed on permanent exhibit in the Procurement Division of the Treasury Department. (M. Price, *Design and Craftsmanship*, 1938, p 26; *Arts and Decoration*, Nov. 1938, p 17)

1939 **Joslyn makes stainless.** Joslyn Manufacturing Company, Fort Wayne, Ind., manufactures its first stainless steel.

1939 **Revere Ware.** Revere Ware copper-bottomed stainless steel cooking utensils are introduced. In 1801, Paul Revere founded the Revere Copper Company, which became the Revere Copper and Brass Company, Inc., the makers of Revere Ware.

1939 **Stainless washer tubs introduced by Speed Queen.** A switch from Monel, a nickel-copper alloy, to stainless steel washer and dryer tubs using type 302 with a No. 3 finish is announced by the Speed Queen Division of McGraw-Edison.

1939 **Brearley receives Sheffield Scroll.** In June 1939, Harry Brearley receives the Freedom of Sheffield Scroll and the Freedom of Sheffield Casket, which is a small metal box decorated with figures depicting the metal trades.

1940 **Alloy Casting Institute (ACI) organized.** The ACI is organized primarily to establish designations for stainless steel casting alloys. E.A. Schoefer, a metallurgical consultant, is the Executive Secretary. The organization is a successor of the Alloy Casting Research Group that was established in 1937.

1940 **Stainless on buildings to be evaluated.** C.C. Snyder of the Republic Steel Company and others from ASTM's Committee A-10 on Stainless Steel make their second, five-year inspection of the stainless steel spire of the Chrysler Building in New York City. Their report, which appeared in the *ASTM Proceedings* of 1940, stated, "The steel on the roof of the Chrysler Building showed no corrosion whatever after ten years."

1940 **Stainless for aircraft valves.** During World War II, chromium stainless steel was used in valves for the engines that powered all British aircraft for the Royal Air Force and the last versions of the American-built Mustang fighter planes.

1940 **Budd receives largest stainless order.** In approximately 1935, the Edward G. Budd Manufacturing Company in Philadelphia sets up a Truck Trailer Division in addition to its Auto Body and Railcar Divisions. An announcement in the April 24, 1940, edition of *Time* magazine reveals that the Fruehauf Trucking Company of Detroit has become a national distributor for the stainless steel trailer sets and has entered an initial order of 10,000 Budd trailers. This represented Budd's largest single order for stainless steel equipment and was a $9 million order. The trailer sets, which are made of approximately two tons of type 301 stainless steel, are shipped disassembled.

1940 **Most stainless alloys in ASTM A 240.** The largest collection of stainless steel alloys in a single specification, ASTM A 240, "Heat-Resisting Chromium and Chromium-Nickel Stainless Steel Plate, Sheet and Strip," is first published.

1940 **American Rolling Mills buys Rustless stock.** American Rolling Mills, Middletown, Ohio, now owns 48.6% of Rustless Iron and Steel Corporation stock. The Baltimore, Maryland, company was organized in 1930.

1940 **AMS specifications published.** In approximately 1940, the Society of Automotive Engineers (SAE) commenced the publication of Aircraft Material Specifications (AMS). These specifications are written by SAE committees representing the manufacturers of military aircraft and include the special requirements of the Army and Navy. The specification numbers consist of four digits, with stainless steels being covered by AMS 53*xx* through 56*xx*. In general, each AMS specification covers a single alloy.

1941 **ACI numbers for stainless casting alloys.** In the December issue of *Metal Progress*, the Alloy Casting Institute (ACI) announces the development of a designation system for "high nickel-chromium and straight chromium alloys used in heat and corrosion resisting castings."

The system is said to provide a concise description of a specific alloy, in contrast to the recent AISI designations for wrought stainless alloys, which convey little information. The ACI designations divide casting alloys into two groups: a "C" series for corrosion-resisting alloys, and an "H" series for heat-resisting alloys. A second letter, from "A" to "Z," is used to denote approximate combined amounts of nickel and chromium. Other details of the system are spelled out in Chapter 16, "The Naming and Numbering of Stainless Steels."

The 1941 article listed 46 designations and their corresponding chemical compositions, which included CA-14 for the least alloyed and HX for the highest alloyed. The C series numbers are used to designate alloys for use below 1200 °F and the 11 designations are for alloys suitable for use above 1200 °F.

1941 **British "En" numbering system is established.** "En," the first British numbering system for steel, was started during World War II. It is believed that "En" stood for emergency number, but this does not seem to have been recorded.

The numbers 56 through 61 were assigned to stainless steels, with letter suffixes. An alloy similar to AISI 304, for example, became En 58E. These early "En" numbers should not be confused with the modern "EN" numbering system, which stands for Euronorm number.

The "En" system became outdated and was replaced in 1967 by a six-digit system that had no name.

1942 **Type 430 wire for voice-recording.** Type 430 stainless steel, a ferritic chromium alloy, is used to make wire 0.004 inch in diameter for voice-recording machines. Thousands of miles of this wire were used for this purpose during World War II.

1942 **Nitrogen added to stainless.** Electro Metallurgical Company, a unit of Union Carbide and Carbon Corporation, announces that small amounts of nitrogen enhance the properties of chromium and chromium-nickel stainless steels.

1943 **18-8 exhaust manifolds.** Solar Turbines, San Diego, Calif., manufactures over 300,000 18-8 stainless steel exhaust manifolds for U.S. planes during World War II.

1943 **Corrosion test methods.** ASTM A 262, "Standard Practices for Detecting Susceptibility to Intergranular Corrosion in Austenitic Stainless Steels," is first published by the American Society for Testing Materials. This standard describes five different tests, including the Strauss and Huey tests.

1943 **Budd builds cargo planes for the military.** In 1942, the Budd Company in Philadelphia sent proposals to the U.S. Army and Navy to build cargo planes, which were in short supply and could not be built because of the wartime shortage of aluminum. The Budd planes would be built of high-strength type 301 stainless steel.

Budd had a fine record of building stainless steel trains, but their experience with aircraft was limited to the building of a tiny four-seat amphibian biplane that flew 1000 hours on demonstration flights in America and Europe in 1931. Nevertheless, the U.S. Navy ordered 200 planes, and the U.S. Army ordered 400. This was particularly remarkable because of Budd's minimal experience with building planes. It is also believed that only sketches of proposed planes were available. A huge new plant was built on Red Lion Road in Northeast Philadelphia to produce the planes. There was attached land on which to build an airfield. The first plane was completed and test-flown on October 27, 1943.

Budd built only 25 of the huge RB-1 Conestoga planes. There were some production problems and a price increase, which are said to have been the reasons that the contract was cancelled. The war had also ended.

Seventeen of the world's only stainless steel cargo planes were bought from Government Surplus for service with the newly established Flying Tigers Air Freight Company. These first stainless steel cargo planes had some unique features, such as the loading ramp that dropped down from the tail, which became the standard design for future military cargo planes.

1943 **Stainless-clad specifications.** ASTM publishes the first two specifications for stainless-clad steel, ASTM A 263, "Corrosion-Resisting Chromium Steel-Clad Plate, Sheet and Strip," and ASTM A 264, "Chromium-Nickel Steel-Clad Plate Sheet and Strip."

1943 **Edward Budd receives ASME medal.** Edward G. Budd, who has been called "the father of the stainless steel streamlined train," receives the highest award of the American Society for Mechanical Engineers (ASME): a medal for "outstanding engineering achievements." Budd received the medal just three years before his death at the age of 76.

1943 **NACE is formed.** The National Association of Corrosion Engineers (NACE) is formed in Houston.

1943 **German stainless grades.** In November 1943, the Verein Deutscher Eisenhüttenleute (VDEh) (now the Steel Institute VDEh) published the first list of all steel grades manufactured at that time in Germany. This list provided the group 4*xxx* for stainless steels.

1944 **Stainless bar specification.** The first ASTM specification for stainless steel bars and shapes, ASTM A 276, is published.

1945 **Passivity.** The phenomenon of passivity is demonstrated by R. Franks (Fig. 15) in the paper "Chromium Steels of Low Carbon Content," which appeared in *ASM Transactions.* Franks exposed iron-chromium alloys, with various amounts of chromium, to corrosion and rusting in an industrial atmosphere for ten years. His graph showed that corrosion was negligible when alloys contained above 10% chromium.

1945 **Precipitation-hardening stainless.** Stainless W, the first commercial heat in the "class of precipitation-hardening stainless steels," is melted by the Carnegie-Illinois Company at Pittsburgh. This class of stainless steels permits a piece to be finish-machined and then given an aging treatment, which hardens the steel, at a temperature low enough to avoid distortion or scaling.

1945 **Stainless watch screws.** The first stainless steel watch screws are made at the Hamilton Watch Company in Lancaster, Pennsylvania. The tiny hardened-and-tempered screws are made on automatic screw machines using type 420F, a free-machining, hardenable stainless steel.

1945 **ASTM specification on iron-chromium-nickel castings is published.** ASTM A 297 on iron-chromium-nickel castings is published.

1945 **Rustless Iron and Steel blackens stainless.** Harold W. Cobb (first cousin of the author), a chemist at the Rustless Iron and Steel Company of Baltimore, Maryland, develops a chemical blackening process for stainless steel. This unusual development was to satisfy a need to eliminate reflection in certain military applications of stainless steel.

1946 **Stainless Steel Fabricators Association.** The Stainless Steel Fabricators Association (SSFA), a forerunner of the British Stainless Steel Association, is formed.

1946 **Budd planes purchased by Flying Tigers.** Seventeen of the Budd Company's RB-1 Conestoga stainless steel cargo planes are purchased as war surplus from the U.S. Navy by the newly formed Flying Tiger line, which was the first scheduled cargo airline in the United States. (See also the 1943 entry on Budd cargo planes.)

1946 **Powder metallurgy stainless parts.** Stainless steel parts made by powder metallurgy appear in the marketplace, according to an article in *Material and Methods*, Vol 23, p 1726.

1947 **Stainless-clad steel plates.** Stainless-clad steel plates are introduced by the Lukens Steel Company, Coatesville, Pennsylvania. Lukens, one of the largest plate mills, was the principal supplier of clad steel plates for many years.

1947 **Stainless spring wire.** ASTM specification A 313 on stainless steel spring wire is published.

1947 **Stainless bars.** ASTM specification A 314 on bars and billets for forgings is published.

1947 **Stainless tubing.** ASTM specification A 213 on austenitic steel tubing is published.

1948 **Armco introduces 17-4 PH.** Armco Steel introduces the first commercially produced family of precipitation-hardening stainless steel alloys, including 17-4 PH, 17-7 PH, and PH 15-7 Mo. This new classification of stainless steels comprises alloys that may be forged and finish-machined, followed by an aging treatment that is performed at a temperature sufficiently low to avoid distortion or scaling of the metal.

1949 **Zapffe's book on stainless steels.** *Stainless Steels* by Carl A. Zapffe is the third book on stainless steel to be published in America. The 350-page book is published by the American Society for Metals, Cleveland, Ohio. Unlike Thum's book, which was a collection of papers written by 75 experts, Zapffe's excellent book was written in the style of a text book and had 20 to 40 questions at the end of each chapter. Because of the chapter on stainless steel history in the book, Zapffe was considered by authorities to be one of the top historians of stainless steel.

The translation of *Stainless Steels* into Japanese is cited as the reason for Japan's proficiency in the stainless steel industry and for their adoption of the AISI numbering system for stainless steels.

1949 **German specification on stainless.** The first German specification for stainless steels, entitled "Nichtrostende Walz- und Schmiedestähle" ("Wrought Stainless Steels"), is published by VDEh as Stahl-Eisen-Werkstoffblatt (Steel Iron Material Sheet) 400 in April 1949.

1949 **Budd rail diesel car (RDC).** From 1949 through 1962, the Budd Company builds the stainless steel RDC, which is a self-propelled diesel-hydraulic passenger car. These cars were primarily adopted for passenger service in rural areas with low traffic density or in short-haul commuter service. The Budd Company built 398 of these cars.

1950 **Highway trailers.** New stainless steel highway trailers, developed by the Budd Company, carry 35% more payload than average vans. The vans, known as Stainless Steel Volume Vans, were introduced by the Fruehauf Trailer Company. The vans had the highest capacity, per length and height, of any trailer in the Fruehauf line. The vans carried up to a 35 ton load.

1950 **AISI *Steel Products Manuals*.** In the 1950s, the American Iron and Steel Institute (AISI) initiates the publication of 16 steel products manuals to cover the major steel mill products. The manuals are intended to provide adjunct material to that included in the ASTM specifications, including data on physical properties, background information, and typical applications. Each of the manuals is to be edited by one of the AISI technical committees, and the manuals are to be revised when deemed necessary.

One of the manuals is *Corrosion- and Heat-Resistant Steels* and is edited by the AISI Technical Committee on Stainless Steel. The editing and publication of these sixteen manuals was turned over to the Iron and Steel Society (I & SS), Warrendale, Pennsylvania, in 1987. The *Corrosion- and Heat-Resistant Steels* manual, with a name change to *Stainless Steels* in 1999, has been through five editions, the latest of which appeared in 2008. It was edited by Harold M. Cobb and published by The Association for Iron and Steel Technology.

1950 **Atlas Steel makes stainless bar.** Atlas Steel Company in Canada manufactures stainless steel in both bar and wire product form.

1950 ***Military Handbook of Metals*.** In approximately 1950, the Department of Defense (DoD) issued MIL-HDBK-H1: *Metals—Cross Index of Chemically Equivalent Metals and Alloys (Ferrous and Nonferrous)*. The book was prepared and revised approximately every three years by the Army Materials and Mechanics Research Center, Watertown, Massachusetts. The book was discontinued when the first edition of *Metals and Alloys in the Unified Numbering System (UNS)* was published in 1975.

1950 **Three-ply cookware produced in Quebec.** In the 1950s, an attractive line of cookware is introduced by the Housewares Division of G.S.W., Ltd., Baie Urie, Quebec. The cookware consists of 18-8 stainless steel inside and out, with a carbon steel core to enhance the distribution of heat. The "Tri-Ply" material was introduced by Allegheny-Ludlum Steel of Brackenridge, Pennsylvania for the cookware market.

1950 **Stainless flatware is imported.** 18-8 stainless steel flatware is imported into the United States from Solingen, a major cutlery center in Germany. The tableware is of fine quality but very simple in design. American silversmiths had not yet begun to produce tableware made of 18-8 stainless steel, but they would soon follow.

1950 to 1960 **Stainless replaces chrome plate for auto applications.** In this decade, there is an increasing use of stainless steel for car trim, such as bumpers, grilles, headlamp surrounds, and wiper arms. The use of chromium-plated steel for these parts is phased out.

1951 **AISI 201 introduced by Allegheny Ludlum.** During the Korean War, the shortage of nickel leads Allegheny Ludlum to develop types 201 and 201L stainless steels, which have similar properties to types 301 and 301L. The manganese content is increased, and the nickel content is decreased. The alloys are austenitic.

1953 **Rolled Alloys Company is formed.** The Rolled Alloys Company is founded at Temperance, Michigan, on the introduction of wrought RA 330 alloy to replace cast HT alloy.

1953 **Roll-formed stainless jet engine compressor blades.** R.B. Wallace & Sons, Silversmiths, of Wallingford, Connecticut, receives a contract from the U.S. Navy to produce stainless steel jet engine compressor blades and vanes using a roll forming method long used in the silverware industry for making knives, forks, and spoons. There was a concern on the part of the U.S. Government that there were not enough drop hammers in the nation to drop forge enough blades for aircraft engines in the event of a national emergency.

The Wallace Aviation Company was established to produce blades on an experimental basis. The results were successful, and it was planned to scale up the operation for production runs. The company was bought by Clevite Corporation of Cleveland, Ohio, in 1955. The company, Clevite Aeroproducts, was in business until 1957, when it was acquired by the Utica

Drop Forge Company, Utica, New York. It was Utica's plan to add the roll forming process as an alternative to their drop forging process. The rolling equipment was moved to the Utica plant.

1954 **Canadian Pacific orders 173 stainless railcars.** The Canadian Pacific Railroad, in an effort to revitalize passenger rail travel across Canada, orders 173 stainless steel railcars from the Budd Company. The cars will equip 18 nine-car, transcontinental trains. The cost is $40 million, or approximately $290 million in 2007 dollars.

1954 **Pullman Standard produces their first all-stainless train.** The Texas Special is the first all-stainless-steel train constructed by Pullman Standard. It is for service on the Missouri, Kansas, & Texas Railroad.

Pullman Standard's entry into the stainless steel train construction business coincides with the expiration date of the Budd patent on the shot weld electric resistance welding process.

1954 **Wallace Silver introduces stainless flatware.** R.B. Wallace & Sons, Silversmiths, of Wallingford, Connecticut, introduces a line of high-quality but very simple 18-8 stainless steel flatware and starts phasing out the production of silver-plated flatware.

1954 **Cleaning and descaling.** ASTM A 380 on the cleaning and descaling of stainless steel parts and equipment is published.

1955 **Allegheny Ludlum melts superalloys by consumable electrode vacuum remelt process.** Allegheny Ludlum is the first to melt superalloys by the consumable electrode vacuum remelt process.

1955 **First DIN stainless specification published.** The first DIN specification for stainless steels is published as Prestandard DIN 17224: 1955-04 with the title "Nichtrostend Stahle für Federn—Güteeigenschafen" ("Stainless Steel for Springs—Quality Properties").

1956 **Socony-Mobil is first stainless steel skyscraper.** The world's first stainless steel skyscraper, the Socony-Mobil Building, a 42-story building, is erected in New York City. It is sheathed with 7000 type 304 stainless steel panels and was the headquarters for the Mobil Oil Corporation from 1956 to 1987.

1956 **Stainless razor blades.** The world's first stainless steel razor blades are introduced by Wilkinson Sword in England.

1957 **Carpenter buys Northeastern Steel.** Carpenter Steel of New

England is formed when Carpenter acquires the bankrupt Northeastern Steel Company at Bridgeport, Conn. The acquisition doubled Carpenter's previous ingot tonnage capacity. A new hot mill was built at Reading, Pennsylvania, and the Bridgeport mill was closed in 1988.

1957 **Cadillac sports stainless roof.** The 1957 Cadillac Eldorado, complete with tail fins, has a shiny stainless steel top.

1957 **J&L establishes Stainless Sheet and Strip Division.** J&L establishes a Stainless Steel Sheet and Strip Division.

1957 **Stainless burnable poison introduced.** In an unusual application, boron carbide is added to type 304 stainless steel to produce a material that will have a high-neutron-absorption cross section for use in reducing the power of a nuclear reactor with a fresh load of fuel. The boron-10 isotope is used instead of boron-11, the naturally occurring isotope, because it has 100 times the neutron-capture capability of boron-11. Commercial production of boron-10 had just begun.

Allegheny Ludlum Steel furnished ⅛ inch diameter wire of this material to the United Nuclear Corporation at New Haven, Conn., for the production of pins for poison plates to be inserted with the fuel loading of nuclear cores for atomic submarines. The capture of neutrons by the boron becomes gradually less as the boron "burns" during the life of a fuel charge and results in a more uniform level of power. The boron is transmuted into helium in the process.

1958 ***Explorer* launched stainless nose cone.** *Explorer*, America's first satellite, is launched at Cape Canaveral. It has a stainless steel nose cone.

1958 **Japan makes stainless railcars.** Stainless steel is adopted in Japan for railway carriages, using U.S. technology and a new high-manganese, low-nickel alloy (16Cr-4Mn-4Ni).

1959 **Budd coach-sleeper cars.** Three new Budd-built Slumbercoaches, the revolutionary coach-sleepers, are acquired by the Baltimore & Ohio Railroad for service between Baltimore-Washington and St. Louis. In designing and building these cars, Budd engineers found a way to provide sleeping room for 40 passengers in the same standard 85 foot long railroad car as the regular sleeper, which has only 22 roomettes. This was done by putting the rooms on two levels, with the uppers two feet above the aisle floor. Certain parts of the upper rooms dovetail with the lower rooms.

1959 **Explosive cladding.** The explosive-cladding process for metals is discovered by the DuPont Company in Wilmington, Delaware. In 1963, DuPont will establish the Detaclad Division to apply this process, bonding stainless steel and other metals.

1959 **Nuclear fuel contained in stainless rods**. The *NS Savannah*, the first nuclear cargo ship, is launched. The nuclear fuel is contained in 3744 stainless steel rods. The ship was decommissioned in 1974 after having steamed more than 450,000 miles.

1959 **Budd's first train, Burlington Zephyr, retires after 3 million miles.** The 25th anniversary of the Budd-built Burlington Zephyr (now the Pioneer Zephyr) stainless steel train is celebrated. The train has traveled over 3 million miles since its famous dawn-to-dusk race from Denver to Chicago. In a 1958 survey conducted by the Institute of Design of the Illinois Institute of Technology, the Pioneer Zephyr was selected by leading designers and architects as one of the "100 best-designed products of modern times."

1960 **Tyson develops Selectaloy system.** In the 1960s, Samuel E. Tyson, a metallurgist at the Carpenter Steel Company, Reading, Pennsylvania, develops the Selectaloy system, which consists of a chart to assist engineers and designers in selecting the right grade of stainless steel for a specific application.

1960 **Sandvik introduces duplex alloy 3RE80.** Sandvik Steel in Sweden introduces duplex alloy 3RE80 in the 1960s. It is particularly designed to resist stress-corrosion cracking.

1960 **Allegheny Ludlum offers bright-annealed stainless.** Allegheny Ludlum is the first company to offer bright-annealed stainless steel in large volume.

1960 **Gillette makes stainless razors.** Gillette safety razors are produced with stainless steel blades.

1960 **Ford makes stainless Thunderbird sports cars in venture with Allegheny Ludlum.** Two stainless steel Thunderbird sports cars are produced by Ford in collaboration with Allegheny Ludlum as a demonstration of the versatility of stainless steel.

1960 **Carpenter patents Custom 450 and Custom 455.** In the 1960s, the Carpenter Steel Company introduces and patents two new precipitation-hardenable stainless steels, Custom 450 and Custom 455, each of which provides a combination of both corrosion resistance and strength in the same alloy.

1960 **Tool and Stainless Steel Association is established.** The Tool and Stainless Steel Industry Association is formed. This group later became the Specialty Steel Industry of the United States

(SSIUS) and, in 1986, became the Specialty Steel Industry of North America (SSINA).

1960 **Report of building inspection for corrosion.** An inspection is made of many of the major buildings in the United States with exterior stainless steel trim or cladding by a group of stainless steel producers of ASTM Committee A-10 on Stainless Steel. This is the sixth inspection over a period of 30 years to determine the condition of the stainless steel surfaces. Buildings were inspected in New York, Philadelphia, Pittsburgh, Cleveland, Chicago, and Florida. There was no instance of corrosion found, except for pitting underneath dirty surfaces. This included the inspection of the Chrysler Building, which had been built in 1930 and was the first building to have stainless steel cladding and ornamentation. It was decided that further inspections would be unnecessary.

1960 **"The Fascinating History of Stainless Steel: The Miracle Metal"**

In 1960, the Republic Steel Company of Cleveland made a movie that skillfully outlined the history of metallurgy and stainless steels and provided an excellent short course in the selection and application of stainless steels. Dr. Carl Zapffe of Baltimore, a consulting metallurgist, created the story and directed the preparation of the illustrations.

Two booklets were made from the movie: Part 1, "Historical Background," and Part 2, "Tailoring Stainless Steel to Do the Job." Five thousand copies were printed of each booklet.

Dr. Zapffe wrote a textbook, *Stainless Steels*, in 1949 and was an avid promoter of stainless steel throughout his career.

1960 **Clad metals.** Composite Metal Products, Inc., Canonsburg, Pennsylvania, is established by John Ulam, who planned to supply bonded metal products to various companies. (See also 1967, clad metal products.)

1961 **Armco introduces Nitronic 40.** Armco Steel introduces the first Nitronic steel, 21-6-9 Stainless (Nitronic 40), a special high-strength, corrosion-resisting alloy strengthened with nitrogen for aircraft hydraulic tubing.

1961 **Budd delivers subway cars.** The Budd Company begins the delivery of 270 stainless steel cars for Philadelphia's SEPTA Market-Frankford elevated line, the backbone of the public transportation system. The cars cost approximately $90,000 each, for a total cost of over $24 million.

1961 **Chrysler spire and tower cleaned.** The Chrysler Building's spire is cleaned, and the stainless steel is found to be in perfect condition after 30 years, despite the aggressive atmosphere due to closeness to the stacks of a coal-fired steam plant.

1961 **Stainless steel publications by ASM.** H. Krainer reports in "50 Years of Stainless Steel," a technical report of Krupp, Vol 20 (No. 4), 1962, that "more than 4,000 treatises were evaluated by the American Society for Metals during the period from 1952 to 1961. During the same period, 898 stainless steel publications were listed by Verein Deutscher Eisenhuettenleute (VDEh)."

1962 **Adjustable safety razor made with types 410, 420, and 416.** The American Safety Razor Company introduces an adjustable safety razor made almost entirely of stainless steel using types 410, 420, and 416.

1962 ***Atlas* rocket is stainless.** On February 20, 1962, John Glenn becomes the first American to orbit the Earth. He was in the *Atlas* rocket, which has been described as a "stainless steel balloon," keeping its shape with the help of a pressurized gas in the interior, also used in pushing out the fuel.

1962 **Disposable stainless needles introduced in England.** Gillette Surgical, Reading, England, begins manufacturing disposable stainless steel hypodermic needles.

1962 **German Werkstoff Numbers for stainless.** The first German Werkstoff Numbers are assigned to German stainless steels. The numbers 1.4000 through 1.4999 were reserved for stainless steels. The numbers are used for both wrought and cast alloys.

1962 **Carpenter improves stainless machinability.** Carpenter Steel develops a method of improving the machinability of type 416 stainless steel by control of chemistry between close limits. It was branded Project 70 stainless, signifying that it had been made available eight years ahead of time. The same improvements were later made for types 303, 304, and 316.

1962 **Centro Inox is formed in Italy.** Centro Inox is founded in 1962. The Italian Stainless Steel Development Association is headquartered in Milan.

1962 **Stainless airport transporter built for Dulles Airport.** The Dulles Airport, 26 miles west of Washington, D.C., is opened. The terminal building was designed by Eero Saarinen of Hamden, Connecticut. The concept that made this airport outstanding from the passenger's viewpoint was that there were no extensions onto the field for aircraft loading.

A specially designed stainless steel mobile transport lounge is used to transport passengers to the aircraft parked on a jet ramp ½ mile away. The mobile lounge was designed by Chrysler in cooperation with the Budd Company. The stainless steel lounge is 54 feet long, 16 feet wide, and 17.5 feet high. It can carry 102 passengers, 71 of them seated. The 35 ton lounge is the largest passenger vehicle to ride on rubber tires. The Budd Company will receive orders for 18 more of these lounges.

1962 **Oneida Silver introduces ornate stainless flatware.** In the early 1960s, the quality and stature of stainless steel flatware permitted the Oneida Silver Company, Oneida, New York, to introduce the first ornate traditional pierced pattern in stainless steel. The rise of stainless steel popularity sparked Oneida's recovery and led the company into a new era of growth.

1962 **Bars and shapes for boilers and pressure vessels.** ASTM A 479 for stainless steel bars and shapes for pressure vessels is published.

1962 **Forgings.** ASTM specification A 473 for stainless steel forgings is published.

1962 **Weaving wire.** ASTM A 478 on stainless steel weaving wire is published.

1963 **Teflon-coated blades introduced by Schick.** Schick becomes the first U.S. manufacturer to produce Teflon-coated stainless steel razor blades.

1963 **Stainless beer kegs.** The first stainless steel beer kegs appear on the U.S. market.

1963 **Atlas acquired by Rio Algoma.** Atlas Steel in Canada is acquired by Rio Algoma.

1963 **Specialty Steel Industry of North America established.** In approximately 1963, the Specialty Steel Industry of North America (SSINA) is established by the Market Development Committee of the American Iron and Steel Institute to promote the use of stainless steel. Over the years, the organization published several dozen pamphlets on the selection of alloys and guidelines regarding the use of stainless steel. As of 2008, these pamphlets may be found on the SSINA website.

1963 **Rope wire.** ASTM A 492 on stainless steel rope wire is published

1963 **United States satellite *Explorer 17* launched.** On April 3, 1963, the National Aeronautics and Space Administration (NASA)

launched a thin-skinned, 35 inch diameter, spherical stainless steel shell containing complex instrumentation. The sphere is made of type 321 stainless steel and has a wall thickness of 0.025 inch. The stainless steel spacecraft structure was designed by the Goddard Space Flight Center and built by the Budd Company.

1964 **Crucible patents Rezistal 422.** Crucible Steel Company receives a patent for type 422 (Rezistal 422) stainless steel, which is a ferritic chromium alloy developed principally for the manufacture of furnace parts that must resist scaling up to 1800 °F (980 °C).

1964 **Budd's largest order is for 600 stainless New York subway cars.** The New York City Transit Authority orders 600 stainless steel subway cars from the Budd Company, Philadelphia, Pa. This is Budd's largest order for rail cars to date. It is estimated that the cost was on the order of $90 million. The cars would have made a train nine miles long. Approximately 4600 tons of type 201 stainless steel were required, the cost of which would be over $29 million in the year 2008.

1964 **SASSDA opens office in South Africa.** The South African Stainless Steel Development Association (SASSDA) is opened in Durban North, South Africa.

1965 **Stainless exhaust system replacements.** Stainless steel replacements for exhaust systems are launched into the U.S. car market.

1965 ***New Horizons in Architecture with Stainless Steel* is published.** The Committee of Stainless Steel Producers of the American Iron and Steel Institute, New York, prepares *New Horizons in Architecture with Stainless Steel.* It is a fine collection of 15 of the most dramatic buildings that include stainless steel in their construction.

1966 **St. Louis Gateway Arch.** The stainless-clad Gateway Arch in St. Louis is completed. Designed by Eero Saarinen, it is the tallest monument in the nation, standing 630 feet high, with a 630 foot span. The monument is clad with 886 tons of ¼ inch thick type 304 stainless steel plates with a No. 3 finish. It is the world's tallest stainless steel monument.

(The Washington Monument is 550 feet high.)

1966 **Allegheny Ludlum patents type 409.** Allegheny Ludlum is awarded a patent for the development of type 409 stainless steel, an 11% chromium ferritic alloy, for automotive exhaust systems.

1966 **Tidal power station.** The world's first tidal power station, near St. Malo, France, is completed with stainless steel turbine blades.

1966 **Age-hardening bars and shapes.** ASTM specification A 564 on hot-rolled and cold-finished age-hardening stainless steel bars and shapes is published.

1967 **British Steel Corporation formed.** Fourteen major U.K. steel companies combine to form the British Steel Corporation (BSC).

1967 **Lincoln convertibles.** Ford Motor Company, in collaboration with Allegheny Ludlum, produces three more stainless steel cars, the 1967 model Lincoln convertibles, to be used by Allegheny Ludlum to promote the use of stainless steel.

1967 **Specification for free-machining bars.** The first ASTM specification on free-machining stainless steel bars, A 582, is published.

1967 **Clad metal products.** Composite Metal Products, Canonsburg, Pa., forms an alliance with Alcoa and becomes Clad Metals, Inc. (See also 1971, clad cookware.)

1967 **Free-machining wire.** ASTM A 581 on free-machining stainless steel wire is published.

1967 **A new alloy numbering system is established in the United Kingdom.** Standard BS 970, "Designation System for Steels," is first issued. The numbers for stainless steel are to conform to BS 970, Part 4, "Stainless, Heat Resisting and Valve Steels." The system replaces the "En" designation system that was in use since World War II.

The series 300–499 is allocated for the first three digits, which are to correspond with AISI stainless steel designations where possible. The letter "S" is the fourth digit and denotes that the alloy falls in the broad classification of stainless steel. The fifth and sixth digits are arbitrarily designated within the range 11 to 99 and are used to designate modifications of the basic alloy chemistry. AISI 304, for example, is assigned the numbers 304S12, 304S14, and 304S15.

1968 **First AOD vessel.** In April 1968, the world's first argon-oxygen decarburization (AOD) vessel for refining stainless steel is placed in operation at the Fort Wayne, Indiana, plant of the Joslyn Manufacturing and Supply Company after 12 years under development. The refining method produces stainless steel with unusually low carbon content, improves overall quality, and reduces cost.

1968 **Electralloy founded.** Electralloy Company is founded at Oil City, Pennsylvania.

1968 **Crucible patents types 416 and 416 Plus X.** Crucible Steel Company receives a patent for type 416 and type 416 Plus X stainless steels.

1968 **Stainless coins circulated in Italy.** A 100 lire coin is introduced in the Italy Republic.

1969 **Concorde flight.** The British supersonic transport, the Concorde, makes its first flight. The plane has a stainless steel rudder, ailerons, and engine nacelles to withstand the relatively high temperature produced at supersonic speeds.

1969 ***Saturn V* rocket.** The first men on the moon (*Apollo 1*) are taken there by a stainless steel *Saturn V* rocket.

1969 **Chicago Transit orders 160 cars.** The Chicago Transit Authority orders 160 Budd-built stainless steel cars.

1970 **ACI merges with SFSA.** The Alloy Casting Institute (ACI) becomes a part of the Steel Founders' Society of America (SFSA), which will continue the ACI research activities and the ACI designations for stainless and nickel alloy castings that were started in 1941.

1970 **Ballpoint refills.** Stainless steel ballpoint pen refills become available.

1970 to 1979 **Home and other applications.** In this decade, there is a marked increase in the popularity of stainless steel in the home environment, including kitchen sinks, electric appliances, kitchen utensils, and washing machine drums. The material also begins to play a more utilitarian role, being used for sanitary ware, street furniture, and stainless steel roofs and building fixtures.

1971 **Avesta 3RE60.** Avesta introduces one of the first duplex stainless steels, Avesta 3RE60.

1971 **Joslyn adds nitrogen to AOD process.** Joslyn Stainless Steels introduces the gaseous nitrogen practice for argon-oxygen decarburization (AOD) refining of stainless steels.

1971 **Armco 18 SR.** Armco Steel introduces 18 SR Stainless, a cost-effective, ferritic chromium stainless steel with excellent oxidation resistance at high temperatures for automobile exhaust systems.

1971 **Clad cookware.** Clad Metals, Inc., Canonsburg, Pennsylvania, begins making stainless-clad cookware under the name All-Clad Metalcrafters. (See also 1988, clad cookware.)

1971 **Crucible type 303 Plus X.** Crucible Steel obtains a patent for the free-machining type 303 Plus X stainless steel.

1972 **ASTM Steel and Stainless Steel Committees merge.** ASTM Committee A-10 on Stainless Steel and Related Alloys merges with ASTM Committee A-1 on Steel to form ASTM Committee A-1 on Steel, Stainless Steel and Related Alloys, creating a committee with approximately 900 members. Committee A-10 had been organized in 1929. The merger was needed principally because specifications including both alloy and stainless steels were required by both committees.

1972 **Specification on annealed or cold-worked stainless sheet.** ASTM A 666, "Annealed or Cold-Worked Austenitic Stainless Steel Strip, Plate, and Flat Bar," is published.

1972 **Major German stainless specification.** The first major German stainless steel specification, "Stainless Steels—Quality Specifications," is published by DIN.

1973 **Stainless magazine started.** *Stainless Steel Industry* magazine is first published. It is the official magazine of the British Stainless Steel Association and is published six times a year. The magazine publishes news items on companies, individuals, new products, and technical developments.

1973 **Armco Nitronic 60.** Armco Steel Corporation introduces Nitronic 60, a nitrogen-bearing stainless steel that is antigalling and wear resistant.

1974 **LTV buys J&L.** J&L becomes a wholly owned subsidiary of the LTV Corporation.

1974 **Colored stainless.** Allegheny Ludlum develops a process to give stainless steel a decorative bronze color, to be used for Western Electric's wall-mounted coin telephone face plates.

1974 **Precipitation-hardening stainless specification.** ASTM specification A 693, "Precipitation-Hardening Stainless and Heat-Resisting Steel Plate, Sheet, and Strip," is published.

1974 **Turkey circulates stainless coins**. Turkey introduces a 25 kurus coin.

1975 **UNS book is published.** The first edition of *Metals and Alloys in the Unified Numbering System (UNS)*, ASTM DS 56, is published jointly by the American Society for Testing and Materials (ASTM) and the Society of Automotive Engineers (SAE). The UNS was established in 1974 for assigning designations to metals and alloys according to a standard system that divides all metals and alloys into 18 series of designations consisting of a letter and five digits. The system includes an "S" series for wrought stainless steels and a "J" series for cast steel and stainless steel.

The book contains a listing of approximately 1000 metals and alloys that have been assigned UNS designations. Type 304L stainless steel is given the UNS designation S30403, alloy XM-1 is S20300, and 17-4 PH is S17400.

The Department of Defense will use DS 56 and discontinue publication of MIL-HDBK-H1: *Metals.*

1975 **Turkey introduces a second stainless coin.** A 50 kurus coin is circulated in Turkey.

1975 **Report of train wreck.** A significant advantage of using austenitic chromium-nickel stainless steel in railcar construction is the unusual extent to which the metal becomes stronger as it is bent, meaning that it can absorb more of the energy of a collision than other materials.

"This superiority of stainless steel in this regard was illustrated in Australia in 1975. A B-Class diesel locomotive drawing carbon steel railcars was coming down a hill at Frankton, southwest of Melbourne. The brakes failed and the locomotive ran into the back of a stationary suburban train sitting in a station at the bottom of the hill. (The stationary train consisted of stainless steel railcars manufactured by Hitachi). The rear stainless steel railcar, which was the first to be hit, ended up sitting on top of the B-Class locomotive, which lifted it up as it ploughed underneath. No one was killed and passengers escaped with minor injuries."

Nickel Magazine, Dec. 1999

1976 **Statue of Liberty.** The Statue of Liberty is inspected during the American bicentennial year to determine the condition of the inner metal skeleton. Dr. Robert Babaoian, a corrosion specialist with the Texas Instruments Company at Attleboro, Mass., climbed up inside the statue to find a badly rusted supporting structure. He arranged to have all of the steel replaced with stainless steel.

1977 **3CR12 stainless alloy developed in South Africa.** The first 25 ton trial heat of 3CR12 ferritic-martensitic chromium stainless steel is cast at Middelburg Steel and Alloys in Middelburg, South Africa. This is the utilitarian, low-cost stainless that will be widely used for applications not requiring a shiny finish, such as for coal cars. The steel has the lowest possible chromium content (10.5%) to be called a stainless steel. (3CR12 is listed as UNS S41003; see also 2003, utility stainless.)

1978 **Uddeholm and Grange-Nyby merger.** Uddeholm's stainless steel operations merge with Grange-Nyby to form Grange-Nyby Uddeholm AB.

1978 **Vacuum bottles.** The world's first stainless steel vacuum bottles are introduced by the Thermos Company to replace glass.

1979 **Intergranular attack in ferritic stainless steels.** ASTM A 763 for detecting intergranular attack in ferritic stainless steels is published.

1979 **Thermostats for car engines.** Stainless steel is used to produce thermostats for car engines.

1980 **First stainless train is national engineering landmark.** The Budd-built Pioneer Zephyr (1934) three-car stainless steel train of the Chicago, Burlington, & Quincy Railroad is designated as a National Historic Mechanical Engineering Landmark by the American Society for Mechanical Engineers (ASME). The train, which was in service for 25 years and covered 3.2 million miles, is on permanent display at the Chicago Museum of Science and Industry.

1980 **Stainless buses predominate in Italy.** Beginning in 1980, Italian buses start using type 304 stainless in construction. The buses are 10% lighter, have a 10% improvement in crash worthiness of the passenger compartment, have reduced maintenance, and are more fuel-efficient. In 2008, 80% of the buses are stainless.

1980 **Thames Flood Barrier.** The massive Thames Flood Barrier makes extensive use of stainless steel in its construction.

1980 to 1989 **Armco aluminized type 409.** During this decade, Armco Steel, Middletown, Ohio, develops aluminized types 409 and 439 stainless steel products, which significantly extend the life of automotive exhaust systems by providing excellent resistance to muffler condensate and pitting from road salt.

1981 **Slater acquires Joslyn.** Slater Steel acquires Joslyn Stainless Steel Company.

1981 **Armco Nitronic 30.** Armco Steel introduces Nitronic 30, a nitrogen-strengthened austenitic stainless steel with resistance to abrasion and metal-to-metal wear.

1981 **Chicago Transit orders 300 cars.** The Chicago Transit Authority orders 300 Budd-built stainless steel cars.

1981 **DeLorean's stainless automobile.** The only automobile to go into production with a stainless steel skin is built by DeLorean

Motor Company in Dunmurry, Ireland. John Z. DeLorean had been one of General Motor's most respected engineers. Over a three-year period, DeLorean built 8563 cars clad with type 304 stainless steel. The company closed because of bankruptcy. Today, the cars are collectors' items.

1982 **U.S. Steel ends stainless production.** In approximately 1982, the U.S. Steel Corporation discontinues the production of stainless steel after being one of the leaders in the business for 55 years.

1982 **Borated steel.** ASTM A 887 for borated stainless steel plate, sheet, and strip for nuclear applications is published. (See also 1957, stainless burnable poison.)

1982 **Japan assumes ISO Steel Secretariat.** The Secretariat of ISO/TC 17 on Steel is relinquished to Japan by the United Kingdom after almost 30 years. The Secretariat will be managed by the Japan Iron and Steel Institute. There are 20 subcommittees in ISO TC 17, including ISO/TC 17/SC 4 on alloy steels and stainless steels. Glenn Selby of the Armco Steel Company, Middletown, Ohio, is the chairman of the U.S. delegation. The International Organization for Standards (ISO) is headquartered in Geneva.

1984 **Mass-produced partial stainless exhaust systems.** The Ford Motor Company mass-produces partial stainless steel exhaust systems. Before the turn of the century, all cars produced in North America will have exhaust systems made completely of stainless steel.

1984 **LTV and Republic merge.** LTV Corporation and Republic Steel Corporation merge to form the LTV Steel Co.

1984 **Chromium Association started.** The International Chromium Development Association (ICDA) is organized. It is headquartered in Paris.

1984 **Allegheny Ludlum AL 29-4.** A patent is issued to Allegheny Ludlum for AL 29-4 alloy to be used in residential furnaces.

1984 **Telegraph poles.** Stainless steel is used for the first time to make telegraph poles.

1984 **Armco 311 DQ.** Armco Steel introduces Armco 311 DQ, an austenitic stainless steel containing copper and nitrogen, to provide higher yield strength and better formability than type 304.

1984 **Avesta AB formed.** Avesta AB is formed through an agreement of Sweden's main stainless steel suppliers: Avesta Jernverks AB, Nyby Uddeholm AB, Fagersta AB, and Sandvik AB.

1984 **Japan enters stainless train business.** From 1984 through 1990, stainless steel is used for half of Japan's passenger trains. Since that time, the stainless steel cars have become the standard for Japanese commuter trains. The cars are said to be 20% lighter than the normal carbon steel cars and 3% lighter than aluminum cars.

1985 **Allegheny Ludlum AL-6XN.** A patent is issued to Allegheny Ludlum for AL-6XN, a nitrogen-bearing, superaustenitic stainless steel developed to minimize chloride pitting.

1985 **Concrete reinforcing.** Thirty-three tons of type 304 concrete reinforcing steel bars are used in the construction of an Interstate Highway I-696 bridge deck near Detroit, Michigan.

1986 **Lloyd's Building.** The Lloyd's Building in London is completed. Lloyd's represents the largest use of stainless steel in a single building in the United Kingdom to this date. "The stainless-clad building has the strongest impact ever. The voluminous, but delicate building has become one of the landmarks of modern London."

1986 **Armco aluminized foil for catalytic converters.** Armco Steel introduces aluminized stainless steel foil for metal monolithic catalytic converters.

1986 **J&L Specialty Products to acquire assets of LTV.** J&L Specialty Products, Inc. is formed to acquire the assets of LTV Steel's Specialty Steel Company.

1986 **Telephone kiosks.** Stainless steel telephone kiosks make their debut in the United Kingdom.

1986 **British specification for rebar.** British standard BS 6744-1986, "Stainless Steel Bars for the Reinforcement and Use of Concrete," is published.

1986 **Armco 301 LN.** Armco Steel introduces 301 LN, a low-carbon, high-nitrogen version of type 301, for high-strength applications where welding is required.

1987 **Budd makes last shipment of cars.** The Budd Company ships the last of 10,641 stainless steel railcars since its first delivery 55 years ago in 1932. The number of cars built would make a train approximately 170 miles long. The amount of stainless steel used was approximately 82,000 tons. Based on the price of stainless steel in 2008, the value of the steel was $524 million. The company closed because of foreign competition and a refusal of the union to agree to a 10% cut in wages. The Budd blueprints for the trains were sold to Bombardier, the railcar builder in Montreal.

1988 **75th anniversary of stainless.** The 75th anniversary of Harry Brearley's discovery of 12% chromium stainless steel, known as cutlery stainless, is celebrated when Sheffield-based British Steel Stainless organizes a full program of national and local events.

1988 **Barraclough paper on history.** "Sheffield and the Development of Stainless Steel" is the title of a paper presented by Dr. K.C. Barraclough at the Brearley Centre, British Steel Stainless, Sheffield, as part of the 75th anniversary celebrations of the discovery of stainless steel by Harry Brearley. The 13-page, illustrated paper is one of the best short histories of stainless steel. It includes 117 references.

1988 **Clad cookware.** All-Clad, the cookware manufacturer of Canonsburg, Pennsylvania, is purchased by Sam Michael, a Pittsburgh entrepreneur, who plans to expand production. (See also 1971, clad cookware.)

1988 **Nickel Development Institute organized.** The Nickel Development Institute (NiDI) is organized. (See also 2004, Nickel Institute.)

1988 **British Steel Stainless created.** British Steel Stainless is created as the dedicated stainless arm of British Steel PLC.

1989 **Brearley autobiography.** *Harry Brearley—Stainless Pioneer* is published by British Steel Stainless in conjunction with the Kelham Island Industrial Museum, Sheffield, to commemorate the 75th anniversary of Brearley's discovery. The 110-page book is a collection of Brearley's notes describing his childhood and business career. The notes were retrieved from Brearley's grandson, who was living in Australia.

1989 **International Molybdenum Association established.** The International Molybdenum Association (IMA) is formed at Pittsburgh, Pennsylvania.

1989 ***Stainless Steel World* magazine.** *Stainless Steel World* magazine is published by KCI Publishing, Zutphen, The Netherlands, as a global magazine for users, suppliers, and fabricators.

1989 **Sammi purchases Atlas.** Sammi Group purchases all Atlas facilities, including the Tracy Quebec plant.

1989 **Ugine acquires J&L Specialty Products.** Ugine, a division of Usinor of France, acquires J&L Specialty Products.

1990 **Armco Nitronic 19D.** Armco Steel introduces Nitronic 19D, a duplex stainless steel casting alloy for automotive structural parts.

1990 **North American Stainless established by Armco and Acerinox.** North American Stainless Company is established as a joint venture between Acerinox and Armco, with headquarters in Ghent, Kentucky.

1990 **Chromium Association.** The International Chromium Development Association (ICDA) is organized, with headquarters in Paris. The organization has 103 members from 26 countries.

1990 ***Corrosion- and Heat-Resisting Steels* is published.** The third edition of this book, which was one of the *Steel Products Manuals* that had been initiated in 1950 by AISI, was revised and edited by members of the AISI Stainless Steel Committee. Wallace Edsel of Allegheny Ludlum, Brackenridge, Pennsylvania, chaired the committee. The book, however, was published by the Iron & Steel Society (I&SS) of Warrendale, Pennsylvania, because AISI turned over the publication of the *Steel Products Manuals* to the I&SS in 1987.

1991 **Electralloy and G.O. Carlson merge.** Electralloy of Oil City, Pennsylvania, merges with the G.O. Carlson Company of Thorndale, Pennsylvania.

1991 **Tallest building in the United Kingdom stainless clad.** The 800 foot Canary Wharf Tower, the tallest building in the United Kingdom, is completed. It is completely clad with stainless steel. Cesar Pelli is the architect.

1992 **Avesta Sheffield formed.** British Steel Stainless and Avesta AB merge to form Avesta Sheffield AB.

1992 **British Stainless Steel Association organized.** The British Stainless Steel Association (BSSA) is organized at Sheffield, England.

1992 **Stainless shipments.** U.S. shipments of stainless steel in 1992 are 1,514,222 tons, according to the U.S. Department of Commerce.

1992 **Australian Stainless Association formed.** The Australian Stainless Steel Association (ASSA) is established, with headquarters in Brisbane.

1994 ***Stainless Steels Handbook* published.** ASM Specialty Handbook *Stainless Steels*, edited by J.R. Davis, is a 575-page book. An unusual feature of this fine reference book on stainless steels is the extensive collection of micrographs provided by George Vander Voort.

1994 **Water storage tank.** The Nickel Development Institute (NiDI) reports what may be the first use of stainless steel to construct a municipal water storage tank. The tank, built at Matsuyama, Japan, is constructed of three grades of stainless steel: S30400, S31600, and S31803. Although more costly to build than the usual concrete tanks, a cost study, including construction costs plus maintenance, indicated that stainless steel was the best option, based on a useful life estimate of at least 60 years.

1994 **Armco and Kawasaki form A-K Steel.** Armco Steel and Kawasaki Steel merge to form A-K Steel, headquartered at Middletown, Ohio.

1994 **60 years of duplex stainless steels.** At Corrosion '94, the Annual Conference and Corrosion Show sponsored by NACE, the paper "60 Years of Duplex Stainless Steel Applications" is presented by Jan Olsson and Mats Liljas of Avesta Sheffield.

1994 **Brazil adopts stainless coinage.** Brazil circulates six denominations of ferritic stainless circular coins, including 1, 5, 20, 25, and 50 centavos and a 1 Real coin.

1995 **Krupp and Thyssen form Thyssen Nirosta.** Krupp and Thyssen merge their stainless steel flat rolled products divisions to form Thyssen Nirosta GmbH, the world's largest producer of stainless steel flat products.

1995 **High-silicon stainless.** In approximately 1995, a new alloy, 700 Si, containing 7% silicon, the highest silicon content in an iron-base alloy, is produced especially for handling hot sulfuric acid. The iron-nickel-chromium-silicon alloy's UNS designation is S70003, and it is the only alloy to have an S7*xxxx* designation.

1995 **Canada recycles 191 railcars.** "Recycling a Railcar Classic" is the title of a Nickel Development Institute (NiDI) report that discusses a decision of VIA Rail, Canada's publicly owned passenger rail company, to refurbish its 40 year old transcontinental fleet of stainless railcars. The initial program involved completely upgrading 191 stainless steel railcars, 187 of which had been supplied to the Canadian Pacific Railway by the Edward G. Budd Manufacturing Company of Philadelphia from 1954 to 1955.

Because the stainless steel siding and roofs are virtually corrosion-free, they require little attention except for washing. The bulk of the restoration will consist of new interior furnishings and a new electrical power supply. The cost per car will be less than $1 million, or approximately half the cost of new cars. The cars are expected to be in service for at least 20 years.

1995 **101 AOD installations** Starting with one argon-oxygen decarburization (AOD) installation for refining stainless steels and other metals in 1970, there are now 101 installations worldwide.

1995 **EN (Euronorm) designations established for steel.** Eighteen countries in Europe will replace their traditional designations for steel and stainless steel with what have been the DIN systems, including a five-digit number and a steel-naming code. For type 304 stainless steel, the number will be 1.4301, and the name will be X5CrNi 18-10. Eighty-three stainless steel alloy compositions and their designations are listed in publication EN 10088-1.

1995 **Chinese develop numbering system for steel.** The Chinese introduce the Iron and Steel Code (ISC), Unified Numbering System for Iron, Steel, and Alloys. The system, modeled after the American Unified Number System (UNS), employs a letter followed by five digits. As in the American UNS, the letter "S" is used for stainless steels.

1995 **Chrysler Building inspected for corrosion.** The Chrysler Building stainless steel dome is inspected for corrosion for the first time since 1965. The building, now 65 years old, represents the longest outdoor exposure test of the metal. Ms. Catherine M. Houska, an architectural consultant for TMR Stainless in Pittsburgh, inspected the tower and gargoyles on behalf of the Nickel Development Institute of Toronto. At the same time, an inspection was made of the tower and spire by Roy Matway of J&L Specialty Metals. The inspections came after a cleaning of the building.

Both inspectors reported that the stainless steel panels are in good condition, except for some pitting found with holes approximately 0.005 inch deep in the 0.020 inch thick panels. "The better-than-expected condition of the stainless steel in a coastal location exposed to high pollution levels during much of the structure's life is attributed to the height of the building which ensures that the top portions of the building are cleaned by wind-blown rain at near hurricane force during heavier rainstorms." Ms. Houska expects that the stainless steel should serve for at least another 100 years.

1995 **James D. Redmond to chair UNS Council.** James Redmond becomes the chairman of the Advisory Council of the Unified Numbering System (UNS) upon the resignation of Alvin G. Cook, retired, Allegheny Ludlum, who chaired the council since 1975. Redmond will also serve as the number assignor for stainless steel alloys. He will also become the editor of *Metals and Alloys in the Unified Numbering System*.

1996 **Harmonized standard compositions.** ASTM A 959, "Standard Guide for Specifying Harmonized Standard Grade Compositions for Wrought Stainless Steels," is published for the first time. This project was headed by Samuel E. Tyson, retired, Carpenter Steel Company, Reading, Pennsylvania.

1996 **PM parts.** Stainless steel powder metallurgy (PM) parts appear in the exhaust systems of U.S. cars. By 2002, almost 60 million PM exhaust system parts will have been produced, using 8000 tons of stainless steel powder. (See also 2002, stainless powders.)

1996 **Exhaust systems.** A large part of the exhaust systems of practically all cars produced in the United States is made from ferritic chromium stainless steel. Approximately 23 kilograms of stainless steel are used per vehicle, which translates to 400,000 tons per year in a marketplace of 15 million vehicles per year.

Research in the past decade on the addition of titanium and niobium has led to greatly improved performance of ferritic stainless steels with 18% chromium (type 439), according to a paper presented at a meeting of the Iron & Steel Society (I&SS), Warrendale, Pennsylvania.

1996 **Reduction of chromite ores**. Armco Steel develops the reduction of chromite and nickel ores with carbon in a rotary hearth to produce feedstock for making stainless steel.

1996 **Armco 410 Cb.** Armco Steel introduces Armco 410 Cb, a heat treatable alloy with high strength and impact resistance for improved exhaust flange applications.

1996 **Stainless Club in Korea.** The Stainless Steel Club is formed in Seoul, Korea, to promote the use of stainless steel.

1996 **International Stainless Steel Forum established.** The International Stainless Steel Forum (ISSF) is founded by the International Iron & Steel Institute in Brussels, Belgium. It is a nonprofit research organization that serves as a world forum on various aspects of the international stainless steel industry. The ISSF comprises 72 company and affiliate members in 26 countries. The national members of ISSF include the following:

- British Stainless Steel Association (www.bssa.org.uk)
- Institut de Developpment de l'Inox (ID Inox) (www.idinox.com)
- CEDINOX (www.cedinox.es)
- Japan Stainless Steel Association (www.jssa.gr.jp)
- CENDI (www.cendi.org.mx)
- Núcleo Inox (www.nucleoinox.org.br)
- Centro Inox (www.centroinox.it)
- Southern Africa Stainless Steel Development Association (SASSDA) (www.sassda.co.za)
- Edelstahl Vereinigung e.V. (www.stahl-online.de)
- Special Steel and Alloys Consumers and Suppliers Association (USSA) (www.ussa.su)
- Euro Inox (www.euro-inox.org)
- Specialty Steel Industry of North America (SSINA) (www.ssina.com)
- Indian Stainless Steel Development Association (ISSDA) (www.stainlessindia.org)
- Stainless Steel Council of China Special Steel Enterprises Association (CSSC) (www.cssc.org.cn)
- Informationsstelle Edelstahl Rostfrei (ISER) (www.edelstahl-rostfrei.de)
- Thai Stainless Steel Development Association (www.tssda.org)

1997 **Type 409 stainless is withdrawn.** In an unprecedented move, ASTM's Committee A-1 on Steel, Stainless Steel, and Related Alloys voted to remove type 409 (UNS S40900) stainless steel from ASTM specification A 240, "Chromium and Chromium-Nickel Stainless Steel Plate, Sheet, and Strip for Pressure Vessels and General Applications," and replace the alloy with three modifications of type 409. The new alloys are stabilized with columbium or titanium, or both, and are identified only according to the UNS designations S40910, S40920, and S40930. These steels are intended for automotive exhaust systems and structural applications where appearance is not a critical requirement.

1997 **Duplex steels book.** The first book on duplex steels, *Duplex Stainless Steels*, is published. The book was edited by Robert N. Gunn and published by Abington. Duplex steels have an attractive combination of properties, including high strength and excellent resistance to chloride stress corrosion.

1997 **Krupp Thyssen Stainless formed.** As part of the reorganization of the Nirosta Division of the Krupp Group, Krupp Thyssen Stainless (KTS) is formed.

1997 **Types 304, 304L, and 316 approved for drinking water systems.** The American National Standards Institute (ANSI)/NSF International standard 61 lists requirements for drinking water system components and deals with contaminants that migrate or extract into the water, including their maximum allowable limits. Stainless steel types 304, 304L, and 316L are approved for such systems.

1997 **Rebar for walls of Guildhall Building.** The Guildhall Building, which has been the center for government in London since the 13th century, is gaining a £50 million addition that uses stainless steel reinforcing bars to ensure its long life. More than 140 tonnes of S30400 stainless steel bars from 6 to 25 millimeters in diameter will be used. It is interesting to note that the reinforcing is intended for the walls of the building, because it is expected that the floor will flex over time as ground conditions change, so that the rigid walls on the upper floors may crack, allowing water to seep in.

1997 **Armco strip pickled with peroxide.** Armco Steel uses hydrogen peroxide to provide clean, pickled stainless steel strip.

1997 **Feltmetal.** Technetics Corp. of Deland, Florida, manufactures Feltmetal acoustic media, an engineered porous product made by the sintering of metal fibers, including types 300 and 400 stainless steel fibers. Technetics acquired Huyck Metals of Milford, Connecticut, the developers of this Feltmetal product in 1962.

1998 **Krupp Thyssen and Pudong to form ShanghaiKrupp Stainless.** Krupp Thyssen Stainless and Shanghai Pudong Iron & Steel agree to build a stainless steel flat rolled products plant in Pudong under the name ShanghaiKrupp Stainless.

1998 **J&L wholly owned by Usinor.** J&L becomes a wholly owned subsidiary of Usinor, a French stainless steel producer.

1998 **Shanghai skyscraper.** The Jin Mao Building is erected in Shanghai. It was designed by the American firm Skidmore, Owens, and Merrill. The 1379 foot building has a stylized stainless steel, aluminum, granite, and glass facade. The building has 88 floors, because 88 means "double good luck." Until 2007, the building was the tallest in the world, based on a roof height of 1214 feet.

1999 **Clad cookware.** Waterford Wedgwood buys All-Clad cookware for $110 million. (See also 2004, clad cookware.)

1999 **Petronas Twin Towers.** The tallest buildings in the world, The Petronas Twin Towers, are clad with 700,000 square feet (65,000 square meters) of type 316 stainless steel. The towers, at 1483 feet (452 meters), are 33 feet higher than the Sears Tower in Chicago.

1999 **Allegheny Technologies formed.** Allegheny Technologies, Inc. is formed, consisting of Allegheny Ludlum, Allvac, Oremet-Wah Chang, Titanium Industries, and Rome Metals.

1999 **Allegheny Technologies acquires Washington Steel.** Allegheny Technologies, Inc. acquires the assets of Lukens' Washington Steel Division from Bethlehem Steel Corporation to enhance its finishing capacity for sheet and strip.

1999 **Thyssen Krupp Materials and Service formed.** Thyssen and Krupp merge to form Thyssen Krupp Materials and Service AG.

1999 **India circulates stainless coins.** Beginning in 1999, a series of six circular ferritic stainless steel coins are introduced, including coins of 1, 2, 5, 10, 25, and 50 rupees.

1999 **Third edition of *Introduction to Stainless Steels.*** The third edition of *Introduction to Stainless Steels*, by J. Beddoes and J.G. Parr, is published by ASM International. The book contains 315 pages.

1999 ***Stainless Steels Product Manual.*** *Stainless Steels*, a *Steel Products Manual* edited by H.M. Cobb, is published by the Iron & Steel Society, Warrendale, Pennsylvania. This is the fourth edition of a reference book originally prepared by the Stainless Steel Committee of the American Iron and Steel Institute (AISI) in approximately 1950. The AISI turned over the revisions and editing of this and 15 other *Steel Products Manuals* to the Iron & Steel Society in 1987. With this edition, the name of the publication is changed from *Corrosion- and Heat-Resisting Steels* to *Stainless Steels*.

This general reference book includes the chemical compositions, mechanical and physical properties, UNS designations, common names and trade names, and typical applications for 225 wrought stainless steel alloys.

1999 ***Stainless Steel & Nickel Alloys* is published.** This handbook is an excellent review of stainless steels and nickel alloys, both cast and wrought. The book is co-edited by Casti Publishing,

Inc. of Edmonton, Alberta, Canada, and ASM International of Materials Park, Ohio. John Bringas, president of Casti, is the Executive Editor, and Stephen Lamb is the Technical Editor.

2000 ***Alloy Digest: Stainless Steels.*** *Stainless Steels*, an *Alloy Digest Sourcebook* edited by J.R. Davis, is published by ASM International, Materials Park, Ohio. The 584-page book provides a collection of 280 data sheets published in *Alloy Digest* by Oscar O. Miller over the past 40 years.

2000 **Stainless sculpture.** "Escaping Flatland" is a large stainless steel sculpture by Edward Tufte. Each of the two units of sculpture consists of four plates of stainless steel that are 2.5 inches thick. The maximum height is 12 feet, and the sculpture covers an area approximately 15 by 30 feet. "Escaping Flatland" reacts wonderfully to changes in the time of day, the season, and the surrounding landscape.

The work is shown in a catalogue for Tufte's show at Artists Space in New York City. The price is listed at $20,000.

2000 **Stainless steel in cars.** The use of stainless steel in cars reaches 65 pounds per car, mainly for the exhaust systems.

2000 **Largest stainless building in North America.** The Edmonton, Canada, composting facility opens. At 23,000 square meters, it is the largest stainless steel building in North America. The siding, roofing, and bolts are made of S30400 stainless steel.

2001 **AvestaPolarit formed.** AvestaPolarit is formed by the merger of the Finnish stainless steel division within Outokumpu, Outokumpu Steel, and the Swedish-British company Avesta Sheffield. British Steel Stainless in Sheffield had been acquired by Avesta. Outokumpu operates the former British Steel Stainless plant at Sheffield.

2001 **China is largest consumer.** China becomes the largest consumer of stainless steel, overtaking the United States and Japan. Stainless steel consumption in China was approximately 2.25 million tonnes; U.S. consumption was approximately 2 million tonnes; and Japan's was approximately 1.5 million tonnes.

2001 ***Pocketbook of Standard Wrought Steels.*** The *Pocketbook of Standard Wrought Steels,* by Harold M. Cobb, is published by the Iron & Steel Society, Warrendale, Pennsylvania. This pocketbook was originally published by AISI in the 1950s and entitled *Pocketbook of Standard Steels.* The purpose was to list the chemical compositions of carbon, alloy, stainless, and tool steels. The 2001 edition is the first edition of the pocketbook to include the nonstandard stainless steels (those types that never were assigned AISI numbers). The chemical compositions for 127 stainless steels are included, along with the type, common name, and UNS designations.

2001 **Nickel-free stainless is introduced.** The Carpenter Steel Company introduces BioDur 128 (UNS S29108) to meet a demand, especially in Europe and parts of Asia, for nickel-free austenitic stainless steels for surgical, dental, jewelry, and food-handling applications. The austenitic structure is maintained by introducing manganese levels of approximately 22%. The high nitrogen level contributes to high levels of strength and corrosion resistance.

2002 **Stainless powders.** The Hoeganaes Company of Cinnaminson, New Jersey, and Electralloy, a G.O. Carlson, Inc. company of Oil City, Pennsylvania, form a partnership to produce high-quality stainless steel powders.

Hoeganaes' new multimillion-dollar water atomization facility at Electralloy's plant is online and brings together the advanced argon-oxygen decarburization melting technology and Hoeganaes' experience in atomizing and powder processing. The new process, with a 10 ton melt size, provides enough capacity to meet the entire requirements of all powder metallurgy applications in North America.

2002 **Outokumpu acquires Avesta Sheffield.** The Finnish firm Outokumpu acquires Avesta Sheffield, the British and Swedish company.

2002 **Arcelor formed.** The world's largest steel company, Arcelor, is created on July 27 through a merger of Acelaria of Spain, Arbed SA of Luxembourg, and Usinor of France. The company has a capacity of 40 million tonnes and is headquartered in Luxembourg.

2002 **Stainless strip continuous casting.** ThyssenKrupp Nirosta GmbH, Krefield, Germany, announces a stainless steel strip casting process that significantly shortens the process compared to continuous slab casting. Hot strip in thicknesses from 1.5 to 4 millimeters is produced by this process.

2002 **J&L is a subsidiary of Arcelor.** J&L Specialty Steel, Inc. becomes a subsidiary of Arcelor Group. J&L is headquartered in Pittsburgh.

2002 **Stainless rebar.** Nearly 300 tons of 2205 duplex alloy stainless steel concrete reinforcing bars are used in a bridge at Coos Bay on the coast of Oregon. The alloy was chosen for its resistance to chloride-induced corrosion and is expected to last for 120 years. It is believed to be the largest installation of stainless steel reinforcing bars in North America.

2002 **Stainless buses offer savings.** Australia's first stainless steel buses are manufactured by Bus Tech Pty Ltd. for Volvo Australia. The exterior and interior skins are made of grade 304 stainless and attached to the stainless steel shell with polyurethane adhesive. The use of stainless steel has resulted in a 700 kilogram reduction in tare weight, a $2 per kilometer savings in petroleum, and the ability for each vehicle to carry an additional 9 passengers.

2002 **Mont Blanc Tunnel repairs.** Mont Blanc Tunnel is reopened after a fatal truck fire in the tunnel in 1999. Safety has been improved through the extensive use of S31603 stainless steel for ventilation fans, lighting equipment, ceiling cladding, piping, fittings, and anchor bolts.

2002 **Corrosion costs.** The Specialty Steel Industry of North America (SSINA) estimates that the cost of corrosion in the United States is $279 billion, or 3.2% of the Gross Domestic Product.

2002 **AgIon antimicrobial coating.** AK Coatings, a wholly owned subsidiary of AK Steel, Middletown, Ohio, introduces flat rolled stainless and carbon steel with the silver-containing AgIon antimicrobial coating. The coating suppresses a broad array of destructive microbes, including bacteria, molds, and fungi. The steel sheet can be used in construction, for food equipment and appliances, and for heating, ventilating, and air conditioning systems.

2003 **Petronas Towers.** At 1242 feet, not counting the antenna, the Petronas Twin Towers buildings in Kuala Lumpur, Malaysia, are the world's tallest buildings, being 33 feet taller than the Sears Tower in Chicago. The buildings are clad with 700,000 square feet of stainless steel.

2003 **ASTM and SAE/AMS specifications.** ASTM, West Conshohocken, Pennsylvania, publishes the first volume to include the complete text of all ASTM and SAE/AMS specifications for stainless steel products. The 2800-page volume is edited by Harold M. Cobb and contains approximately 400 specifications. The ASTM specifications are taken from 13 volumes of the *ASTM Book of Standards*.

2003 **Stainless bridge.** A 142 meter stainless steel bridge is completed at Bilbao, Spain. Avesta Sheffield supplied 500 tonnes of high-strength, duplex grade SAF 2304 (1.4462) for the project.

2003 **British guide to stainless specifications.** The *Guide to Stainless Steel Specifications* is published by the British Stainless Steel Association (BSSA). The book uses only the European EN designations for stainless steels in lieu of the former British designations, such as 304S17. The EN designations are based on the German Werkstoff Numbering system. There is considerable cross referencing to ASTM specifications. The EN designations are not to be confused with the former U.K. En numbers.

2003 **Utility stainless.** 3CR12 (S41003) is being used in Australia, India, and South Africa to good advantage and cost-effectively for stainless steel wagons for hauling large tonnages of coal. Because the 3CR12 alloy is smooth and hard, the discharges of the coal from the wagons is greatly improved over that of carbon steel wagons, where shaking and hammering are necessary to dislodge the coal.

2003 **Stainless rivets.** In Singapore, 9 million aluminum rivets in the windows of 43,000 residential apartments are being replaced with S30400 stainless steel. The aluminum rivets that were installed in the aluminum casement windows between 1987 and 1998 had become loose.

2004 **Clad cookware.** Groupe SEB, Sweden, buys All-Clad cookware from Waterford Wedgwood, with the intent to market the product worldwide. The All-Clad plant remains in Canonsburg, Pennsylvania.

2004 **10th edition of *Metals and Alloys in the Unified Numbering System*.** The 10th edition of *Metals and Alloys in the Unified Numbering System (UNS)* is jointly published by SAE and ASTM.

2004 **Nickel Institute organized.** The Nickel Institute is established through a merger of the Nickel Development Institute (NiDI) and the Nickel Producers Environmental Research Association. The nonprofit organization represents the interests of 24 companies that produce more than 90% of the world's annual nickel output.

2004 **Pipe for fire protection and hot and cold water.** The 101-story Taipei Financial Centre uses S30400 and S31600 stainless steel pipe for fire protection and hot and cold water.

2005 **75 year old Chrysler Building is fine.** The stainless steel cladding on the tower of the Chrysler Building in New York City is reported to be in fine condition by those cleaning the outside of the building.

2005 **The Chrysler at 75.** On May 26, 2005, the *New York Times* devotes a 14-page section to the history of the Chrysler Building at 42nd Street and Lexington Avenue in New York City. It is undoubtedly the greatest public recognition of, and tribute to, this building, which was the first where stainless steel was used.

Twenty-four writers contribute articles on various aspects of the building. The first page has the most detailed photograph ever published of the stainless steel panels on the tower. The picture shows a closeup of the top 18 feet of the tower and spire, with stainless steel covering everything except the windows.

2005 ***Stainless Steel Focus, Ltd.* organized.** A magazine published at Boston, United Kingdom, will provide information for the stainless steel industry, including news and analysis of the markets for stainless steels and the raw materials for stainless steel production.

2006 **Zhangjiagang Posco production.** Zhangjiagang Posco Stainless Steel in China produces 1.9 million tons per year of stainless steel. This Korean-owned mill has been producing stainless steel since 1997.

2007 **Budd Commerce Center.** The former 75-acre property of the Budd Company at Hunting Park and Wissahickon Avenues in Philadelphia is now the Budd Commerce Center. There are 20 buildings, offering 2.4 million square feet of space, for commercial and residential development.

The Budd Company's operations were finished here in 2002 and consolidated in Detroit.

2007 **Leading flat stainless products producer.** The world's leading manufacturer of stainless flat products is now ThyssenKrupp Nirosta, together with its affiliated companies ThyssenKrupp Accai Speciali Terni, ThyssenKrupp Mexinox, and Shanghai Krupp Stainless.

2007 **Stainless production at 30 million tons**. The world production of stainless steel is reported to have doubled in the last 10 years to reach 30 million tons in 2007.

2007 **Naming and numbering stainless steels.** The article "The Naming and Numbering of Stainless Steels," by Harold M. Cobb, is published in the September issue of *Advanced Materials & Processes* by ASM International.

2008 **Stainless steel manual published.** *Stainless Steels, A Steel Products Manual*, edited by H.M. Cobb, is published by the Association for Iron and Steel Technology (AIST). The 350-page book is the fifth edition of a book originally compiled by the Stainless Steel Committee of the American Iron and Steel Institute in 1955.

This edition includes the descriptions, chemical compositions, physical and mechanical properties, and applications of 150 stainless steels, as well as the titles of all ASTM and AMS specifications for stainless steels. The book includes an alloy index that lists 2400 trade names and numbers (domestic and foreign) for stainless steels and their equivalent Unified Numbering System (UNS) designations. A new chapter discusses the history of the naming and numbering of stainless steels.

2008 **1936 stainless steel Ford on exhibit.** A Ford Tudor automobile, a strikingly beautiful stainless steel car, was one of four commissioned in 1936 by the Allegheny Steel Company, Pittsburgh, Pennsylvania. It is exhibited at an antique auto show in Louisville, Kentucky. The stainless steel car body was stamped and assembled by the E.G. Budd Manufacturing Company of Philadelphia and completed at the Ford Motor Company. Allegheny Steel used the cars to promote their line of stainless steels. The cars were driven by the Allegheny top stainless steel salesmen each year. The cars were driven for approximately 200,000 miles before being consigned to museums.

2008 **11th edition of *Metals and Alloys in the Unified Number System (UNS)* is published.** The 11th edition of the UNS book is published jointly by ASTM and SAE.

2008 **Hadron Collider to use stainless.** Conseil Européen pour la Recherche Nucléaire (CERN) in Geneva builds the largest particle accelerator in the world. A stainless steel ring is made of 450 tons of Nirosta 4307 (304L) stainless steel.

2008 ***Stainless Steel for Designers.*** This book, by Michael F. McGuire, has a fine discussion of applications for various stainless steels. It is published by ASM International, Materials Park, Ohio.

2008 **Price of stainless.** The price of 18-8 stainless steel sheet in June 2008 reaches a high of $3.30 per pound. The high cost is largely due to the high price of nickel.

2008 **Alloy cross-reference database.** The *International Metallic Materials Cross-Reference Database* is available on diskette from Genium Publishing. This database was originally prepared by metallurgists John G. Gensure and Daniel L. Potts of the General Electric Research Laboratory at Schenectady, New York. A third edition was published in 1987. The authors compared the chemical compositions of 45,000 alloy designations and grouped them according to the Unified Numbering System designations, using their best engineering judgment when the chemistries were not so close as to be obvious.

The book includes cast iron, steel, stainless steel, tool steel, aluminum, copper, and nickel alloy designations.

2008 ***Stainless Steel World News* is online.** Market and trading news and product information is available online.

2008 **Precipitation-hardenable stainless application.** The Lockheed-Martin Joint Strike Fighter is the first aircraft to use a precipitation-hardenable stainless steel in its airframe, according to Carpenter Steel, the supplier of the Custom 465 alloy.

2009 ***Stainless Steel World* interviews author.** The September issue of *Stainless Steel World*, the international stainless steel trade magazine published in AJ Zuphen, The Netherlands, publishes a two-page interview with Harold M. Cobb. The interview focuses on Cobb's association with stainless steel during his 66 year metallurgical career.

A considerable amount of his industrial career was devoted to the development of new stainless steel products. Starting with watch screws at the Hamilton Watch Company, he went on to the building of large stainless steel airplane propeller blades and cold roll-formed compressor blades that could replace forgings for jet engines. He worked on stainless steel laced with boron carbide, which was used as a moderating material (called a burnable poison) for nuclear submarine reactors, and on the development of porous sintered stainless steel fiber structures, known as Feltmetal, to be used for noise abatement in jet engines.

His later career included the editing of books on steel and stainless steel, participation in the development of the Unified Numbering System (UNS) for metals and alloys, and finally, authoring the first book on the history of stainless steel, published in 2010 by ASM International.

2009 **SSINA posts membership.** The Specialty Steel Industry of North America (SSINA) is a voluntary trade association representing virtually all of the producers of specialty steel in North America. The members include ATI Allegheny Ludlum, Pittsburgh, Pennsylvania; Haynes International Inc., Kokomo, Indiana; ThyssenKrupp VDM USA, Inc., Florham Park, New Jersey; ATI Allvac, Monroe, North Carolina; Latrobe Specialty Steel Company, Latrobe, Pennsylvania; Talley Metals Technology, Inc., Hartsville, South Carolina; Carpenter Technology Corporation, Reading, Pennsylvania; Mexinox S.A. de C.V. San Luis Potosi, S.L.F., Mexico; Universal Stainless and Alloy Products, Inc., Bridgeville, Pennsylvania; Crucible Special Metals, Syracuse, New York; North American Stainless, Inc., East Ghent, Kentucky; Valbruna Slater Stainless, Inc., Fort Wayne, Indiana; Electralloy, Oil City, Pennsylvania; and Outokumpu Stainless, Inc., Schaumberg, Illinois.

2009 **Colossal stainless sculpture rises in Mongolia.** In an article on August 2, 2009, the *New York Times* reports the erection of a colossal statue of Genghis Khan, the legendary horsemen who conquered the known world in the 13th century. The 131 foot tall giant on horseback is wrapped with 250 tons of stainless steel. It rests on top of a two-story circular building that serves as a visitors' center. The statue is said to be the largest in the world.

2009 **Rolled Alloys acquires Weir Materials.** Weir Materials in the United Kingdom, noted for their Zeron 100 alloy, is acquired by Rolled Alloys of Temperance, Mich. The new company will be called RA Materials.

2009 **International Iron and Steel Institute changes name.** The International Iron and Steel Institute becomes the World Steel Association (www.worldsteel.org).

2009 **75th anniversary of the Burlington Zephyr.** The first production model of a stainless steel train entered service on May 25, 1934, at which time the three-car streamlined train broke all speed records during a nonstop, dawn-to-dusk race from Denver to Chicago.

The train, which weighed only ⅓ that of a regular Pullman train, used only $17 worth of diesel fuel on the trip—less than two cents per mile. The train, built by the Edward G. Budd Manufacturing Company of Philadelphia, Pennsylvania, was the first of thousands to be built by the company and represented one of the major uses of stainless steel—approximately 100,000 tons over a 50-year period.

An article, "The 75th Anniversary of the Burlington Zephyr Stainless Steel Train," authored by Harold M. Cobb, is published by ASM International in the June 2009 issue of *Advanced Materials & Processes.*

2009 **Flying Yankee celebration.** The Flying Yankee Restoration Group hosts the fourth annual event for those supporting the cause of the train's restoration on April 3, 2009, in Manchester, N.H. The date is the 74th anniversary of the Flying Yankee, the three-car train built by the Edward G. Budd Company, Philadelphia, in 1935.

2009 **Bombardier of Canada to build stainless trains in India.** Bombardier, Ltd. of Montreal will set up facilities in India to meet the growing demand for stainless steel trains in India.

2009 **Stainless steel demand skyrockets in India.** A 55% increase in the use of flat stainless steel products occurs between 2004 and 2010. The 818,000 tonnes used in 2004 is expected to increase to 1,269,000 tons in 2010.

2012 **Flying Yankee restoration.** The restoration of the Flying Yankee three-car train by the Flying Yankee Restoration Group, Inc. is scheduled to be complete in approximately 2012. The Flying Yankee, which was built in 1935, is the second of the streamlined stainless steel trains to be built by the Edward G. Budd Manufacturing Company. After 22 years of service on the Boston & Maine Railroad, the train was retired from service because of insufficient riders.

The train will run short excursions from Lincoln, New Hampshire, in the summer months. In the winter, the train will be at Concord, New Hampshire, for use as a rolling classroom for lessons on history and technology.

The other early stainless steel train of the same design is the Burlington Zephyr, which was built in 1934 and may be seen at the Chicago Museum of Science and Industry.

2012 **100th anniversary of the discovery of chromium-nickel stainless steel.** This year marks the 100th anniversary of the discovery of the commercial use of chromium-nickel stainless steel by Dr. Eduard Maurer and Benno Strauss of the Krupp Steel Works.

2013 **100th anniversary of the discovery of chromium stainless steel.** The year 2013 marks the 100th anniversary of the discovery of the commercial use of chromium stainless steel by Harry Brearley of the Firth Brown Research Laboratories at Sheffield and Elwood Haynes at Kokomo, Indiana.

Index

B

C

D

H

I

P

Q

R

S

T

U

V

W

Z